Field Guide to the
Broad-Leaved Herbaceous Plants

Field Guide to the
Broad-Leaved Herbaceous Plants
of South Texas
Used By Livestock and Wildlife

WITHDRAWN

James H. Everitt
D. Lynn Drawe
Robert I. Lonard

Texas Tech University Press

This book was set in News Gothic and Letraset Compacta. The paper used in this book meets the minimum requirements of ANSI/NISO Z39.48-1992 (R1997).∞

Design by Melissa Bartz

Printed in China

Library of Congress Cataloging-in-Publication Data

Everitt, J. H.
    Field guide to the broad-leaved herbaceous plants of south Texas: used by livestock and wildlife / James H. Everitt, D. Lynn Drawe, Robert I. Lonard.
        p. cm.
    Includes bibliographical references and index.
    ISBN 0-89672-400-X (alk. paper)
    1. Forage plants—Texas, South—Identification. 2. Forbs—Texas, South—Identification. 3. Herbivores–food–Texas, South.
    I. Drawe, Dale Lynn, 1942-   . II. Lonard, Robert I. III. Title.
    SB193.3.U5E94   1999
    633.2'009764—dc21   98-44054
                                                       CIP

99 00 01 02 03 04 05 06 07/ 9 8 7 6 5 4 3 2 1

Texas Tech University Press
Box 41037
Lubbock, Texas 79409-1037 USA

800-832-4042
TTUP@TTU.EDU
Http://www.ttup.ttu.edu

CONTENTS

The senior author dedicates this book to his wife,
Jeanne Lavon Everitt.

PREFACE

This publication contains color photographs, family names, scientific names, and ecological characteristics of broad-leaved, herbaceous plants (forbs) that occur in southern Texas. Keys are provided for the families, genera, and species covered. Grasses and grass-like plants (sedges) are not included here.

Our purpose is not to present another accounting of the wildflowers of the region, although one will recognize many of these plants as attractive roadside residents. We have attempted to select, photograph, and describe plants that are of importance to wildlife, livestock, and man, pointing out the reasons or characteristics for their importance. Although our coverage is not complete, the most commonly-encountered plants have been included. The complete name history (synonymy) for all species is not included here, but in some cases this has been included to reflect recent or persistent usage.

Although this publication focuses on plants that occur in southern Texas, the extensive ranges of many of the represented species make it a useful reference for plants in other areas of Texas, the southwestern United States, and northern Mexico.

Some 185 species, encompassing 143 genera and 51 families are represented. Most of the plants represented are native to the region, but some introduced species are also included. A native plant is indigenous to an area and grows without cultivation, whereas an introduced plant was brought in from another region and reproduces itself in the area. All plants included here are represented by herbarium specimens kept at the Department of Biology (PAUH), University of Texas-Pan American, Edinburg, Texas, and the Rob and Bessie Welder Wildlife Foundation, Sinton, Texas.

This publication will be useful to ranchers and range managers, natural resource and wildlife management personnel, scientists, and anyone interested in the flora of southern Texas. It will enable one to identify many of the herbaceous plant species in this region. Information provided on the ecological characteristics of the plants represented will be useful to help develop sound land management programs.

Further information on the broad-leaved, herbaceous plant species of this area can be found in Donovan S. Correll and Marshall C. Johnston's 1970 *Manual of the Vascular Plants of Texas*; Fred B. Jones' 1975 *Flora of the Texas Coastal Bend*; and Alfred Richardson's 1990 *Plants of Southernmost Texas* and 1995 *Plants of the Rio Grande Delta*.

ACKNOWLEDGEMENTS

The authors thank Rene Davis for his help in obtaining photographs and Jeanne Everitt for her encouragement and assistance in obtaining photographs. Thanks are also extended to Wayne Swanson for preparation of figures and Angie Cardoza for help in word processing. We express our gratitude to Dr. Alfred Richardson, University of Texas at Brownsville, Brownsville, Texas, Dr. A. Michael Powell, Sul Ross State University, Alpine, Texas, and to an anonymous reviewer for reviewing the manuscript.

This publication was partially funded by a grant from the Rob and Bessie Welder Wildlife Foundation and is Welder Wildlife Foundation Contribution Number *B-15.*

INTRODUCTION

The South Texas or Rio Grande Plains and the Gulf or Coastal Prairies and Marshes vegetational areas of southern Texas (Gould, 1975) (see map) are represented by a diverse and complex flora. The rich variety of plant life in this region is attributed to a wide variability in climate, soils, and topography. These factors also create a wide variety of habitats.

The northern portion of south Texas is influenced by a temperate climate and is represented by many plant species that occur in other areas of Texas and the United States. The southern portion of the area is influenced by a sub-tropical climate that allows a number of tropical species from Mexico and Central America to reach this area. Plants of the Gulf Prairies and Marshes occur in the eastern portion of the area. In the western portion of south Texas, conditions become more xeric, permitting a number of species that occur in the Chihuahuan Desert of Mexico to enter the area.

Although woody plants (shrubs and small trees) and cacti are the most conspicuous components of the south Texas vegetational complex, the flora consists primarily of herbaceous species (grasses and broad-leaved, flowering herbs). Broad-leaved, herbaceous plants (forbs) are a category of plants that are often overlooked and undervalued by the landowner. These plants include those commonly referred to as wildflowers and weeds. The wildlife manager or naturalist recognizes these plants as one of the most important groups in the plant world because of their value to the animal community, both the mammalian and avian faunas.

This book is intended to be a guide for identifying broad-leaved, herbaceous plant species that are of importance to wildlife, livestock, and man in south Texas. In addition to the usual presentation of information about each plant, each description includes (where appropriate) a comment on the *value* of each species.

Many forbs have exceptional nutritional value to herbivorous animals; some provide nutrients not available elsewhere in the plant kingdom. Many forbs are exceptional seed producers. Many provide herbage that is highly digestible to the herbivore, unlike the herbage of woody plants and grasses in most growth stages.

Ecologically, a forb can be one of the first seed plants in early successional stages, i.e., an ephemeral annual, or it can be a component of a climax community. In other than early seral stages, however, forbs are usually in the minority in the composition of a grassland community or grassland/brushland complex. Perennial forbs, particularly legumes, are highly desirable in the climax grassland community because they add much-needed variety, and, in the case of legumes, they add nitrogen to the soil.

PLANT NAMES, DESCRIPTIONS, AND GEOGRAPHICAL RANGES

Scientific names for plant species used herein are according to S. L. Hatch, K. N. Gandhi, and L. E. Brown's 1990 *Checklist of the Vascular Plants of Texas*. Common names were obtained from various sources including Correll and Johnston, Jones, Hatch et al., and Richardson. All plant descriptions were based upon observations obtained from freshly collected material or from specimens housed in research facilities listed above. Further, information on plant descriptions and geographical ranges of plant species was obtained from Correll and Johnston, Jones, and Richardson. Comments concerning wildlife, livestock, and human importance were taken from numerous research studies and the authors' personal knowledge.

PLANT IDENTIFICATION AND THE USE OF KEYS

When a herbaceous plant is collected for identification, as much of the plant should be removed as possible. Use a shovel or trowel to carefully remove roots and possible rhizomes. Always collect plants that have flowers and/or fruits. It is virtually impossible to identify plants unless these crucially-important reproductive structures are present (Lonard, 1993).

You will need magnification of at least 10X to identify herbaceous plants. A magnification glass, hand lens, or stereoscopic dissecting microscope, two dissecting needles, and a metric ruler are the only tools required.

The keys that follow represent an artificial device for identifying herbaceous plant families that are important to wildlife and livestock in southern Texas. Successive choices between contrasting statements are followed until the correct family name is found by the process of elimination. Keying herbaceous plants requires skill and practice. One should not guess when keying specimens. Guessing will almost always result in an incorrect identification. Refer to the glossary when you encounter an unfamiliar term. By using the key, it should be possible to identify the specimen's family. You will proceed to the family in the text. What follows will be a key to the genera and species in that family. Located in the text will be a brief species description followed by comments about the species use by wildlife or livestock. After you have reached a tentative identification, examine the photograph of the specimen.

KEY TO MAJOR GROUPS

1a. Plants lacking seeds, reproducing by spores borne in small, nut-like underground sporocarps; leaves resembling a 4-leaf clover. Division: Polypodiophyta (ferns). **Marsileaceae**
1b. Plants reproducing by seeds; seeds borne in fleshy or dry fruits. Division: **Magnoliophyta** (angiosperms) **2**

2a. Leaves usually parallel-veined (sometimes pinnate or palmate); flower parts usually in 3s or 6s or their multiples. Class: **Liliopsida, Key A** (monocotyledons)
2b. Leaves usually pinnately or palmately veined; flower parts usually in 4s or 5s or their multiples (sometimes 2 or more floral parts).Class: **Magnoliopsida, Key B** (dicotyledons)

KEY TO FAMILIES

KEY A: Class Liliopsida (monocotyledons)

1a. Flowers imperfect; plants monoecious. **2**
1b. Flowers perfect; plants with bisexual flowers. **3**

2a. Leaves sagittate; inflorescence a raceme with usually 3 flowers per node. **Alismataceae**
2b. Leaves linear or strap-shaped; inflorescence a dense, interrupted spike. **Typhaceae**

3a. Ovary superior. **4**
3b. Ovary inferior. **5**

4a. Plants with a bulb; perianth of tepals. **Liliaceae**
4b. Plants with fibrous or fleshy roots; bulbs absent; perianth differentiated into a calyx and corolla. **Commelinaceae**

5a. Tepals white on the inner surface; stamens 6. **Amaryllidaceae**
5b. Tepals blue and yellow or dark purple on the inner surface; stamens 3. **Iridaceae**

KEY TO FAMILIES

KEY B: Class Magnoliopsida (dicotyledons)

1a. Plants aquatic; leaves subpeltate, deeply cleft, floating on the surface. **Nymphaeaceae**

1b. Plants on dry land or on muddy shorelines, but most of the plant present above the water; leaves not subpeltate. **2**

2a. Ovary superior. **KEY C**

2b. Ovary inferior (partially inferior in *Portulaca oleracea*) **3**

3a. Tendrils present, low-growing vines; flowers unisexual, plants dioecious; fruit a globose berry. **Cucurbitaceae**

3b. Tendrils absent, if low-growing, not vines; fruit not as above. **4**

4a. Inflorescence a compact head of multiple flowers subtended by phyllaries; calyx often modified into a pappus. **Asteraceae** (Compositae)

4b. Plants not as above. **4**

5a. Inflorescence an umbel or a cone-like head; fruit a schizocarp of 2 mericarps. **Apiaceae** (Umbelliferae)

5b. Inflorescence not as above; fruit a capsule. **5**

6a. Leaves opposite; stipular bristles present. **Rubiaceae**

6b. Leaves alternate (or if opposite); stipular bristles absent. **6**

7a. Petals 4. **Onagraceae**

7b. Petals 5. **8**

8a. Petals yellow or red; sepals 2; shoots succulent. **Portulacaceae**

8b. Petals blue; sepals 5; shoots not succulent. **Campanulaceae**

KEY C: Dicots with a superior ovary

1a. Flowers bisexual. **KEY B**

1b. Flowers unisexual; plants dioecious, monoecious or polygamous. **2**

2a. Perianth of a well-defined calyx and corolla; fruit a drupe. **Menispermaceae**

2b. Perianth reduced to 1 series (only the calyx present; it is often petaloid), or not clearly differentiated into a calyx and corolla. **3**

3a. Fruit a 3-lobed capsule. **Euphorbiaceae**

3b. Fruit an achene. **4**

4a. Plant a vine; stamens numerous; achenes with a persistent, plumose style. *Clematis* in **Ranunculaceae**

4b. Plant not a vine; stamens 4 or 5; achenes not as above. **5**

5a. Sepals 4; stamens 4. **Urticaceae**

5b. Sepals 5; stamens 5. *Amaranthus* in **Amaranthaceae**

KEY D: Dicots with a superior ovary; flowers bisexual

1a. Corolla present. **KEY E**

1b. Corolla absent (can appear petaloid in several groups). **2**

2a. Leaves bipinnately compound; inflorescence a globose head; stamen filaments can be brightly colored; fruit a legume *Mimosa* or *Schrankia* in **Fabaceae**

2b. Plants not having all of the above characteristics. **3**

3a. Leaves compound (trifoliolate) or deeply divided, mostly basal with an involucre above; sepals petaloid. *Anemone* in **Ranunculaceae**

3b. Leaves simple. **4**

4a. Leaves alternate. **5**

4b. Leaves opposite or whorled. **6**

5a. Fruit a small, red berry; stamens 4. **Phytolaccaceae**

5b. Fruit an achene; stamens 6 or more; sepals petaloid. **Polygonaceae**

6a. Fruit a capsule; sepals petaloid. **Aizoaceae**

6b. Fruit a utricle or anthocarp. **7**

7a. Calyx tubular; stamens exserted beyond the calyx lobes; calyx petaloid. **Nyctaginaceae**

7b. Calyx not tubular; stamens not exserted; spiny bracts at base of flowers. **Amaranthaceae**

KEY E: Dicots with a superior ovary; flowers bisexual; corolla present

1a. Leaves compound and alternate or opposite, or simple and alternate. **KEY F**

1b. Leaves simple and opposite, whorled or in a basal rosette **2**

2a. Petals 4, united. **3**

2b. Petals 5 (6), free or united. **4**

KEY TO FAMILIES

3a. Leaves in a basal rosette; inflorescence a spike. **Plantaginaceae**
3b. Leaves opposite or whorled; flowers solitary in leaf axils.
   Loganiaceae

4a. Petals free; corolla actinomorphic. **5**
4b. Petals united; corolla actinomorphic or zygomorphic. **6**

5a. Blades palmately lobed, venation palmate. **Geraniaceae**
5b. Blades with entire margins, venation pinnate. **Caryophyllaceae**

6a. Corolla highly zygomorphic, bilabiate. **Lamiaceae** (Labiatae)
6b. Corolla actinomorphic or only slightly zygomorphic. **7**

7a. Plant a vine; shoots with a milky latex; fruit a follicle.
   Asclepiadaceae
7b. Plants not as above; fruit a capsule or of 4 bony nutlets. **8**

8a. Fruit separating into 4 bony nutlets. **Verbenaceae**
8b. Fruit a capsule. **9**

9a. Calyx lobes 10; stamens 2. **Oleaceae**
9b. Calyx lobes 5; stamens 4 or 5. **10**

10a. Stamens 4; plants perennials; corolla slightly zygomorphic.
   Acanthaceae
10b. Stamens 5; plants annuals; corolla actinomorphic. **Polemoniaceae**

KEY F: Dicots with a superior ovary; flowers bisexual; corolla present; leaves compound and alternate or opposite, or leaves simple and alternate

1a. Leaves simple, alternate. **KEY G**
1b. Leaves compound, alternate or opposite. **2**

2a. Leaves opposite, even-pinnately compound. **Zygophyllaceae**
2b. Leaves alternate. **3**

3a. Leaves bipinnately or tripinnately compound. **4**
3b. Leaves pinnately compound, trifoliolate, or palmately compound. **5**

4a. Fruit a legume; stamens 10. **Fabaceae** (Leguminosae)
4b. Fruit a berry; stamens 5. *Ampelopsis* in **Vitaceae**

KEY TO FAMILIES

5a. Fruit a legume. **Fabaceae** (Leguminosae)
5b. Fruit a capsule. **6**

6a. Sepals 2; petals 4; corolla highly zygomorphic; leaves pinnately dissected. **Fumariaceae**
6b. Sepals 5; petals 5; corolla actinomorphic; leaves trifoliolate. **Oxalidaceae**

KEY G: Dicots with a superior ovary; flowers bisexual; corolla present; leaves simple and alternate

1a. Petals 4. **2**
1b. Petals 5. **3**

2a. Stamens 6, usually 4 long and 2 short; fruit a capsule. **Brassicaceae** (Cruciferae)
2b. Stamens 4, of nearly equal length; fruit a black or purple berry; tendrils usually present. *Cissus* in **Vitaceae**

3a. Stamens united by their filaments. **4**
3b. Stamens free; filaments distinct. **6**

4a. Fruit a legume; stamens usually 10, diadelphous. **Fabaceae** (Leguminosae)
4b. Fruit usually a capsule. **5**

5a. Stamens numerous, 10 or more, monadelphous and forming a hollow tube around the style. **Malvaceae**
5b. Stamens 5. **Sterculiaceae**

6a. Fruit a 1-seeded, prickly pod; stamens 4, free or adnate to the upper petal. **Krameriaceae**
6b. Fruit not as above. **7**

7a. Sepals 2-3; plants spiny or prickly; with a yellow latex. **Papaveraceae**
7b. Sepals 5; plants not as above. **8**

8a. Stamens 10. **Geraniaceae**
8b. Stamens 4, 5, or 6. **9**

9a. Stamens 4 or 6. **10**
9b. Stamens 5. **11**

10a. Corolla actinomorphic; stamens 6, free. **Lythraceae**
10b. Corolla highly zygomorphic; stamens 4, epipetalous.
   Scrophulariaceae

11a. Corolla highly zygomorphic. **Violaceae**
11b. Corolla actinomorphic. **12**

12a. Petals free; stamens free; style 1, branches 5. **Linaceae**
12b. Petals united; stamens adnate to the corolla; style not as above. **13**

13a. Style 1, unbranched; fruit a berry or capsule. **14**
13b. Style 1 or 2, each branched into 2 segments near the apex. **15**

14a. Plants a trailing or climbing vine; fruit a capsule. *Ipomoea* and
   Convolvulus in **Convolvulaceae**
14b. Plants usually not trailing or climbing, but if so, then fruit a berry.
   Solanaceae

15a. Plants annuals; flowers usually in scorpioid cymes, or 1-few in the
   leaf axils. **Hydrophyllaceae**
15b. Plants perennials; flowers usually borne singly. **Convolvulaceae**

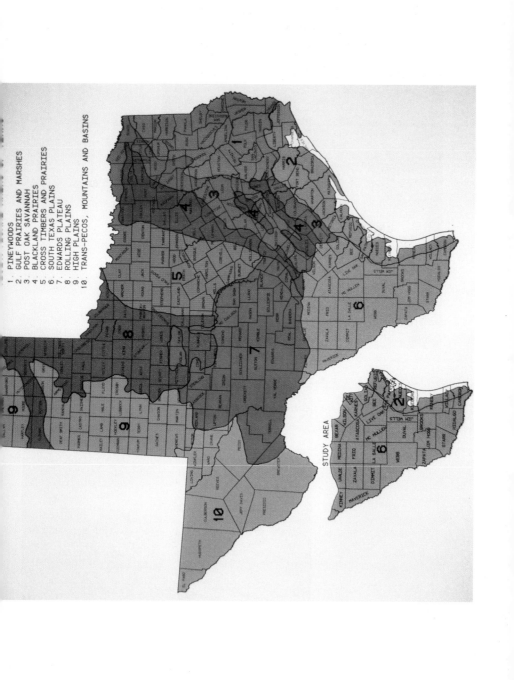

1. PINEYWOODS
2. GULF PRAIRIES AND MARSHES
3. POST OAK SAVANNAH
4. BLACKLAND PRAIRIES
5. CROSS TIMBERS AND PRAIRIES
6. SOUTH TEXAS PLAINS
7. EDWARDS PLATEAU
8. ROLLING PLAINS
9. HIGH PLAINS
10. TRANS-PECOS, MOUNTAINS AND BASINS

STUDY AREA

ACANTHACEAE

Ruellia L.
1a. Corolla 4.5 cm long or longer. ***Ruellia nudiflora***
1b. Corolla about 3 cm long. ***Ruellia runyonii***

RUELLIA, WILD PETUNIA
Ruellia nudiflora (Engelm. & Gray) Urban

PERENNIAL: Erect.
STEMS: Slightly angled, with scattered pubescence.
LEAVES: Simple, opposite, pubescent; blades ovate; margins wavy to entire; apex obtuse or rounded; petioles elongated.
INFLORESCENCE: Few-flowered clusters.
    CALYX: Sepals 5, united, glandular pubescent.
    COROLLA: Petals 5, united, forming a tube below, slightly zygomorphic, blue to purple, 4.5 cm long or longer.
    STAMENS: 4, epipetalous.
    PISTIL: Ovary superior.
FRUIT: An oblanceolate, pubescent capsule.
COMMENTS: Frequent on loam or clay soils in prairies and open brushy areas in the Coastal Prairies and Rio Grande Plains. The leaves are eaten by white-tailed deer and the seeds are eaten by bobwhite quail.

RUELLIA
Ruellia runyonii Tharp & Barkl.

PERENNIAL: From a rhizome.
STEMS: Branching at the upper nodes, with scattered pubescence.
LEAVES: Simple, opposite, with scattered hairs; blades lanceolate to nearly ovate; margins remotely toothed; petioles pubescent.
INFLORESCENCE: Panicles present in the upper nodes; peduncles minutely-glandular pubescent.
    CALYX: Sepals 5, united, lobes linear, glandular pubescent.
    COROLLA: Petals 5, united, slightly zygomorphic, bluish-purple, glandular-pubescent on the outside, about 3 cm long.
    STAMENS: 4, epipetalous, with 2 longer than the others, included within the corolla.
    PISTIL: Ovary superior, pubescent; style 1, unbranched.
FRUIT: A brown, minutely-pubescent capsule; seeds round, flat, with a translucent, white margin.
COMMENTS: Frequent in arroyos, creeks, and river bottoms in the Rio Grande Plains and Coastal Prairies. The leaves are consumed by white-tailed deer and the seeds are eaten by bobwhite quail.

AIZOACEAE

1a. Leaves whorled; stamens 3–4. *Mollugo*
1b. Leaves opposite; stamens about 50; shoots succulent. *Sesuvium*

GREEN CARPETWEED, INDIAN CHICKWEED
Mollugo verticillata L.

ANNUAL: With a shallow taproot.
STEMS: Prostrate, branching freely, forming mats, glabrous.
LEAVES: Simple, whorled, glabrous; blades oblanceolate, linear, or spatulate; margins entire; petioles present.
INFLORESCENCE: Flowers in a cluster or whorl in the leaf axils; pedicels minutely pubescent.
    CALYX: Sepals 5, free, 3-veined on the back, white on the inner surface, petaloid, glabrous.
    COROLLA: Petals absent.
    STAMENS: 3 or 4; anthers and filaments white.
    PISTIL: Ovary superior, glabrous; styles 3.
FRUIT: A many-seeded capsule; seeds ridged, brown.
COMMENTS: Common on sandy and sandy loam soils in openings, prairies, ditches and fields throughout the Rio Grande Plains and Coastal Prairies. The seeds are eaten by bobwhite quail and mourning doves.

SEA PURSLANE, CENICILLA
Sesuvium portulacastrum (L.) L.
PERENNIAL: Rooting at the nodes.
STEMS: Trailing or suberect, branching freely, succulent, green, pink, or reddish, glabrous.
LEAVES: Simple, opposite, glabrous, succulent; blades oblong-oblanceolate, elliptic, ovate, or spatulate; margins entire; petioles with a winged, hyaline margin, nearly ensheathing the stems.
INFLORESCENCE: Flowers solitary, axillary; pedicels glabrous.
    CALYX: Sepals 5, petaloid, pinkish-purple, with a horn-like appendage on the back near the apex.
    COROLLA: Petals absent.
    STAMENS: About 50; perigynous; filaments pink or reddish-purple.
    PISTIL: Ovary superior (perigynous); styles numerous, united near the base; stigmas 3–5.
FRUIT: A circumscissile capsule partially enclosed by the persistent calyx; seeds black, lustrous.
COMMENTS: Common in beach sands and around bay areas in coastal south Texas. The leaves are eaten by jackrabbits.

ALISMATACEAE

LONGLOBE ARROWHEAD
Sagittaria longiloba Engelm.

PERENNIAL: With white rhizomes and vermiform, fibrous roots.
STEMS: Erect, 4-angled, glabrous; spongy with abundant arenchyma tissue.
LEAVES: Simple, in a basal cluster, glabrous; blades sagittate; venation parallel; margins entire; petioles elongated; irregularly angled.
INFLORESCENCE: A raceme with usually 3 flowers per node; pistillate flowers below the staminate; nodes widely spaced, bearing scaly bracts.
STAMINATE FLOWERS
    CALYX: Sepals 3, free.
    COROLLA: Petals 3, free, white, actinomorphic, rounded at the apex.
    STAMENS: More than 10; anthers yellow.
PISTILLATE FLOWERS
    CALYX: Similar to the staminate flowers.
    COROLLA: Similar to the staminate flowers.
    PISTIL: Ovary superior; pistils numerous per flower.
FRUIT: A curved, beaked achene.
COMMENTS: In shallow water or wet places in the Rio Grande Plains and Coastal Prairies. This plant is eaten by sandhill cranes and several species of waterfowl.

AMARANTHACEAE

1a. Leaves alternate; flowers unisexual; plants dioecious. *Amaranthus*
1b. Leaves opposite; flowers bisexual. *Froelichia*

CARELESSWEED, QUELITE, PIGWEED
Amaranthus palmeri S. Wats.

ANNUAL: From a taproot.
STEMS: Often turning red in older plants; glabrous.
LEAVES: Simple, alternate, glabrous; blades lanceolate to ovate; margins entire; petioles green to reddish-green.
INFLORESCENCE: Plants dioecious; flowers in terminal spicate racemes and in axillary clusters below.
STAMINATE FLOWERS
   CALYX: Bracts subtending calyx green, with acuminate apices; sepals 5, free, with a green midvein.
   COROLLA: Absent.
   STAMENS: 5; anthers greenish-yellow.
PISTILLATE FLOWERS
   CALYX: Bracts subtending calyx about twice the length of sepals, with long, acuminate apices; sepals 5, free, green in the midsection; margins scarious; apex mucronate.
   COROLLA: Absent.
   PISTIL: Ovary superior; style branches 2 or occasionally 3.
FRUIT: A utricle with a single, shiny seed.
COMMENTS: On silty, sandy, and gravelly soils in a variety of habitats throughout south Texas. The young foliage is eaten by white-tailed deer, jackrabbits, and livestock, and the seeds are consumed by bobwhite quail, mourning doves, white-winged doves, and Rio Grande turkeys.

AMARANTHACEAE

DRUMMOND SNAKECOTTON
Froelichia drummondii Moq.

PERENNIAL: From a woody rootstock; occasionally an annual.
STEMS: Erect, leafless above, with a dense covering of woolly hairs.
LEAVES: Simple, opposite, subsessile, stipulate, shaggy to woolly
pubescent; blades lanceolate, grayish; margins slightly crisped or entire.
INFLORESCENCE: Flowers in a mass of cottony hairs, in interrupted
clusters near the stem apex.
    CALYX: Bracts subtending calyx 2, lobed, papery, brownish, densely
    pubescent with white, cottony hairs below; sepals 5, lobes linear above.
    COROLLA: Absent.
    STAMENS: 5, united below.
    PISTIL: Ovary superior; style 1.
FRUIT: A 1-seeded utricle.
COMMENTS: In dry sandy soils throughout southern Texas. The seeds are
eaten by bobwhite quail.

AMARYLLIDACEAE

RAINLILY, CEBOLLETA
Cooperia drummondii Herb.

PERENNIAL: From a scaly bulb; flowering after summer rains.
STEMS: A bulb.
LEAVES: Simple, in a basal cluster, venation parallel; blades linear, less
than 0.5 cm wide.
INFLORESCENCE: Flowers solitary on a pedicel up to 30 cm long.
    CALYX and COROLLA: Perianth of 6, united, white tepals; fragrant;
    tepals pink or purple-tinged on the back, actinomorphic; floral tube 10
    cm or more long.
    STAMENS: 6, epipetalous, attached near the base of tepal lobes, white;
    filaments minute; anthers white.
    PISTIL: Ovary inferior, 3-lobed; style 1, as long as the floral tube.
FRUIT: A many-seeded capsule; seeds numerous, black.
COMMENTS: Common on sandy-based soils (following rain) in various
habitats in the Rio Grande Plains and Coastal Prairies. The foliage is eaten
by white-tailed deer.

APIACEAE [Umbelliferae]

APIACEAE (Umbelliferae)
1a. Leaf segments spine-tipped; inflorescence a conelike head. *Eryngium*
1b. Leaf segments not spine-tipped; inflorescence a compound umbel. **2**

2a. Shoots densely pubescent; inflorescence terminal; fruits with glandular-tipped bristles and smaller white hairs. *Daucus*
2b. Shoots glabrous; inflorescence arising from leaf axils; fruits glabrous. *Ciclospermum*

SLIM LOBE CELERY
Ciclospermum leptophyllum (Pers.) Sprague.
[Syn. *Apium leptophyllum* (Pers.) F.V. Muell.]

ANNUAL: From a taproot.
STEMS: Glabrous.
LEAVES: Compound with parsley-like leaves, occasionally 2 per node, glabrous; leaflets linear; petioles slightly grooved, sheathing the stem.
INFLORESCENCE: Flowers in compound umbels arising from the leaf axils.
  CALYX: Sepals inconspicuous, closely adhering to the ovary.
  COROLLA: Petals 5, inserted at apex of ovary, about 0.1–0.3 mm long.
  STAMENS: 5.
  PISTIL: Ovary inferior, glabrous, with longitudinal ribs.
FRUIT: A schizocarp of 2 equal mericarps, rounded, about 1.5 mm long.
COMMENTS: Common in moist or wet soil in the Rio Grande Plains and Coastal Prairies. The leaves are occasionally eaten by white-tailed deer.

SOUTHWESTERN CARROT, RATTLESNAKE WEED
Daucus pusillus Michx.

ANNUAL: From an elongated taproot.
STEMS: Erect, pubescent, with papillose-based hairs.
LEAVES: Bipinnately compound (parsley-like), alternate, pubescent; leaflets about 1.0 mm wide; rachis revolute-grooved.
INFLORESCENCE: Flowers in a terminal, compound umbel; peduncle slightly pubescent, its base subtended by pinnately dissected bracts; pedicels subtended by linear, simple, pubescent bracts.
  CALYX: Sepals inconspicuous or absent.
  COROLLA: Petals 5, free, less than 1.0 mm long.
  STAMENS: 5, free; anthers white.
  PISTIL: Ovary inferior; styles 2.
FRUIT: A schizocarp with glandular-tipped bristles on the ribs and small, white hairs between the ribs.
COMMENTS: Common on a variety of soil types in disturbed areas in the Rio Grande Plains and Coastal Prairies. The leaves are eaten by white-tailed deer.

APIACEAE (Umbelliferae)

Eryngium L.
1a. Inflorescence bracts and sepals purple-tinged. *Eryngium hookeri*
1b. Inflorescence bracts and sepals white. *Eryngium nasturtiifolium*

HOOKER ERYNGO
Eryngium hookeri Walp.

ANNUAL: From a taproot.
STEMS: Erect, glabrous, with vertical ribs.
LEAVES: Simple, usually alternate below and opposite above, glabrous; blades dissected with linear, spine-tipped segments; margins ribbed; petioles sheath-like.
INFLORESCENCE: A conelike head subtended by an involucre of spine-tipped, light purple bracts; flowers subtended by a spine-tipped bract; margins membranous.
    CALYX: Sepals 5, free, apex subulate, purple-tinged.
    COROLLA: Petals absent.
    STAMENS: 5, free.
    PISTIL: Ovary inferior with white, vermiform scales; style 1, unbranched.
FRUIT: A schizocarp of 2 mericarps.
COMMENTS: In moist soils in the Coastal Prairies. The young foliage is eaten by white-tailed deer.

NASTURTIUM LEAF ERYNGO, HIERBA DEL SAPA
Eryngium nasturtiifolium Juss. ex Delar

ANNUAL: From a shallow root system.
STEMS: Glabrous, prostrate to ascending.
LEAVES: Simple, alternate, glabrous; blades lanceolate to oblanceolate; margins lobed, often minutely spine-tipped.
INFLORESCENCE: A conelike head (resembles a "mini-pineapple"), from leaf axils; heads subtended by stout, spine-tipped bracts.
    CALYX: Sepals 5, free; margins white; apex mucronate; subtended by a series of soft, white, scaly, mucronate-tipped bracts.
    COROLLA: Petals 5, free, minute, less than 0.5 mm long.
    STAMENS: 5, free; anthers light brown.
    PISTIL: Ovary inferior; styles 2, unbranched.
FRUIT: A schizocarp with ribs bearing soft spines.
COMMENTS: In moist heavy soil of the Rio Grande Plains. The young foliage is eaten by white-tailed deer and cattle.

ASCLEPIADACEAE

BEARDED SWALLOW-WORT
Cynanchum barbigerum (Scheele) Shinners
[Syn. *Metastelma barbigerum* Scheele]

PERENNIAL: Vine with climbing and twining stems; white latex present.
STEMS: Glabrous; somewhat woody below.
LEAVES: Simple, opposite, glabrous; blades linear-lanceolate to lanceolate; margins entire; apex acuminate; petioles present.
INFLORESCENCE: Few-flowered, umbellate clusters in the leaf axils.
    CALYX: Sepals 5, united near base.
    COROLLA: Petals 5, united, greenish-white, densely pilose within.
    STAMENS: 5, free nearly to the base.
    PISTIL: Ovary superior; style 1, unbranched.
FRUIT: A tapered, glabrous follicle, 3.5–4.0 cm long.
COMMENTS: Common in open woodlands and slopes in the Rio Grande Plains and Coastal Prairies. The leaves are an important food of white-tailed deer and are occasionally eaten by bobwhite quail and Rio Grande turkeys.

ASTERACEAE (Compositae)

1a. Heads ligulate; shoots with a milky latex. **2**
1b. Heads discoid or radiate. **3**

2a. Stems glabrous; leaves with auricles at the base of the petioles.
    Sonchus
2b. Stems with scattered, matted hairs; auricles absent. ***Pyrrhopappus***

3a. Heads discoid. **KEY A**
3b. Heads radiate. **KEY B**

KEY A: Heads discoid

    1a. Leaf margins armed with sharp spines; involucre cup-shaped; phyllaries armed with sharp spines. ***Cirsium***
    1b. Leaf margins lacking sharp spines; involucre not as above. **2**

    2a. Flowers green, staminate and pistillate and in different or same heads; plants with a strong odor. **3**
    2b. Flowers variously colored, but not green, bisexual in each head. **4**

ASTERACEAE [Compositae]

3a. Staminate and pistillate flowers in different heads (monoecious); staminate flowers in a terminal inflorescence; leaves pinnatifid. *Ambrosia*

3b. Staminate flowers in the same head with the pistillate; leaf margins toothed, but not deeply lobed. *Iva*

4a. Plants perennials; leaves linear. *Liatris*

4b. Plants annuals; leaves lanceolate or oblanceolate. **5**

5a. Corollas blue, fragrant; plants usually scandent; pappus of bristles. *Eupatorium*

5b. Corollas lavender; plants erect; pappus of scales. *Palafoxia*

KEY B: Heads radiate

1a. Corollas of ray florets white. **2**

1b. Corollas of ray florets yellow or with combinations of yellow and red. **7**

2a. Corollas of disk florets white. **3**

2b. Corollas of disk florets yellow. **4**

3a. Stems with broad, lateral wings; ray florets conspicuous. *Verbesina*

3b. Stems lacking lateral wings; ray florets about 0.5 mm long. *Parthenium*

4a. Leaves opposite; pappus bractlike with a recurved spine. *Melampodium*

4b. Leaves alternate or absent. **5**

5a. Leaves absent or awl-shaped and spiny. *Leucosyris*

5b. Leaves present, not awl-shaped nor spiny. **6**

6a. Stems glabrous or nearly so; plants erect. *Aster*

6b. Stems densely pubescent; plants low-growing. *Aphanostephus*

7a. Corollas of ray flowers with a combination of red and yellow, sometimes red nearly throughout. **8**

7b. Corollas of ray florets yellow throughout. **12**

8a. Receptacle prominently cone-shaped or hemispherical. **9**

8b. Receptacle neither cone-shaped nor hemispherical. **10**

ASTERACEAE (Compositae)

9a. Plants annuals; corollas of ray florets red only near the base; achenes 4-angled. *Rudbeckia*

9b. Plants perennials; corollas of ray florets red or burgundy cover much of the surface; achenes flattened. *Ratibida*

10a. Corollas of ray florets red well above the base; leaves simple. *Gaillardia*

10b. Corollas of ray florets reddish only at the base (red spot may not be visible until floret is removed); leaves compound. **11**

11a. Leaf segments threadlike, less than 1 mm wide. *Thelesperma*

11b. Leaf segments ovate or linear, but wider than the above. *Coreopsis*

12a. Leaves compound; upper phyllary lobes with paired, linear glands. *Thymophylla*

12b. Leaves simple; phyllaries not as above. **13**

13a. Leaves opposite or basal and opposite above. **14**

13b. Leaves alternate. **17**

14a. Stems glabrous, glaucous; ray florets pistillate. *Helianthus*

14b. Stems pubescent; ray florets bisexual. **15**

15a. Phyllaries 4 (bud-like). *Tetragonotheca*

15b. Phyllaries numerous, more than 4. **16**

16a. Leaves lanceolate, strigose with pustulate-based hairs; turning black when drying. *Wedelia*

16b. Leaves triangular and ovate, pubescence not as above. *Simsia*

17a. Plants perennials. **18**

17b. Plants annuals. **20**

18a. Chaff present; leaves pinnatifid; disk florets not maturing achenes. *Engelmannia*

18b. Chaff absent; leaves entire or margins irregularly lobed; disk florets maturing achenes. **19**

19a. Shoots white, woolly pubescent; pappus of awns in a single series. *Psilostrophe*

19b. Shoots canescent, not white, woolly; pappus of bristles in 2 series, the outer ring inconspicuous. *Heterotheca*

20a. Chaff absent. **21**

20b. Chaff present. **22**

21a. Leaves lanceolate or oblong; shoots pubescent; pappus of spine-tipped scales. **Ambylolepis**
21b. Leaves linear; shoots glabrous, but glutinous; pappus minute, scale-like. **Gutierrezia**

22a. Phyllaries usually recurved at the apex; pappus of scales and bristles. **Xanthisma**
22b. Phyllaries usually erect, not recurved; pappus of deciduous, scale-like awns or of 2 bristles. **23**

23a. Pappus of 2 deciduous, scalelike awns. **Helianthus**
23b. Pappus of 2 bristles; leaves grayish-green. **Verbesina**

ASTERACEAE [Compositae]

PERENNIAL RAGWEED, WESTERN RAGWEED
Ambrosia cumanensis Kunth in H.B.K.
[Syn. *Ambrosia psilostachya* DC.]

PERENNIAL: With slender lateral roots that resemble rhizomes; forming large colonies; shoots with a strong odor.
STEMS: Erect, hirsute with appressed hairs.
LEAVES: Simple, opposite, hirsute; blades lanceolate to ovate-lanceolate; margins deeply lobed, pinnatifid; petioles hirsute.
INFLORESCENCE: Flowers unisexual; plants monoecious; flowers in inconspicuous heads; males located above females.
  STAMINATE FLOWERS: In racemose aggregations; phyllaries united into an inverted, pubescent, cuplike structure; pollen produced in minute, greenish florets.
  PISTILLATE FLOWERS: Flowers inconspicuous in a cluster of green bracts or phyllaries; pappus absent; fruit an achene.
COMMENTS: Common on a variety of soil types in the Rio Grande Plains and Coastal Prairies. The leaves are an important food of white-tailed deer, javelina, Rio Grande turkeys and chachalaca, and the seeds are eaten by bobwhite quail and mourning doves.

ASTERACEAE (Compositae)

HUISACHE DAISY
Amblyolepis setigera DC.
[Syn. *Helenium setigerum* (DC.) Britt. & Rusby]

ANNUAL: From a taproot; shoots and flowers fragrant.
STEMS: Densely pubescent with long, matted hairs.
LEAVES: Simple, alternate, pubescent; blades oblong below, lanceolate above; uppermost leaves sessile, the lowermost with petioles.
INVOLUCRE: Phyllaries light green, acuminate at the apex, pubescent; peduncle pubescent with papillose-based hairs.
RECEPTACLE: Cone-shaped; heads radiate.
RAY FLORETS: Corollas yellow, 3-lobed at the apex, pistillate; pappus of translucent, spine-tipped scales; ovary densely pubescent.
DISK FLORETS: Corollas yellow; pappus of translucent scales, each with a soft awn at apex; ovary densely pubescent.
FRUIT: A silky, pubescent achene.
COMMENTS: Frequent on sandy loam or caliche in prairies and open areas in the eastern portion of the Rio Grande Plains and Coastal Prairies. The leaves are occasionally eaten by white-tailed deer and cattle.

ASTERACEAE (Compositae)

Aphanostephus DC.
1a. Heads, including florets, 1.0–1.5 cm wide. *Aphanostephus ramosissimus*
1b. Heads, including florets, up to 3 cm wide. *Aphanostephus skirrhobasis*

ARKANSAS LAZY DAISY
Aphanostephus skirrhobasis (DC.) Trel.

ANNUAL: From a taproot.
STEMS: Densely pubescent.
LEAVES: Simple, alternate, pubescent; blades mostly lanceolate to oblanceolate; margins toothed or lobed on the lowermost leaves and entire on the uppermost; petioles present on lowermost leaves, absent on the uppermost.
INVOLUCRE: Phyllaries in 2–3 rows, lanceolate, pubescent.
RECEPTACLE: Chaff absent; heads radiate, up to 3 cm wide.
RAY FLORETS: Corollas white, but pinkish-white on the back, pistillate, swollen below.
DISK FLORETS: Corollas yellow, swollen below; pappus of scale-like awns.
FRUIT: An achene.
COMMENTS: Common on sandy soils in the Rio Grande Plains and Coastal Prairies. The leaves and flowers are frequently eaten by white-tailed deer. This plant is also eaten by cattle.

Aphanostephus skirrhobasis (DC.) Trel. var. *kidderi* B.L. Turner lacks a swollen area at the base of the disk corollas. [Not illustrated.]

Aphanostephus ramosissimus DC. **PLAINS LAZY DAISY**. Similar to *A. skirrhobasis*. However, the heads including florets are only 1.0–1.5 cm wide. This species is common on sandy loam and caliche soils in the Rio Grande Plains. The leaves are an important food of white-tailed deer. [Not illustrated.]

34

ASTERACEAE (Compositae)

Aster L.
1a. Plants annuals; stems slightly pubescent. *Aster subulatus*
1b. Plants perennials with rhizomes; stems glabrous. *Aster ericoides*

HEATH ASTER
Aster ericoides L.

PERENNIAL: With elongated rhizomes, forming colonies.
STEMS: Erect, glabrous.
LEAVES: Simple, alternate, occasionally ciliate; blades sessile, linear; margins entire; apex with a mucronate tip on some leaves; leaves subtending inflorescence 2–4 mm long; leaves on lower stems to 1.2 cm long.
INVOLUCRE: Phyllaries in 3–4 rows, linear-lanceolate; margins ciliate.
RECEPTACLE: Heads radiate.
RAY FLORETS: Corollas white, fertile, pappus of numerous bristles.
DISK FLORETS: Corollas yellow, fertile, pappus similar to the rays, bristles with antrorsely arranged barbs.
FRUIT: An achene.
COMMENTS: Frequent on loam, clay, or caliche soils in the northern part of the Rio Grande Plains and Coastal Prairies. The leaves are eaten by white-tailed deer.

HIERBA DEL MARRANO, PRAIRIE ASTER
Aster subulatus Michx. var. *ligulatus* Shinners

ANNUAL: From a taproot; tall, spindly; continues to flower in lawns after mowing.
STEMS: Slightly pubescent.
LEAVES: Simple, alternate, glabrous; blades narrowly to broadly lanceolate, sessile, with a conspicuous midvein; margins entire; apex acute or acuminate.
INVOLUCRE: Phyllaries in 5–6 rows, glabrous, apex acute.
RECEPTACLE: Naked; heads radiate, about 1 cm wide or more.
RAY FLORETS: Corollas white, turning lavender with age, rolling back from the apex, glabrous; pappus of fine, capillary bristles.
DISK FLORETS: Corollas yellow; pappus similar to the rays.
FRUIT: An achene covered with small bristles.
COMMENTS: Abundant in ditches, swales, margins of ponds, and other poorly drained areas in the Rio Grande Plains and Coastal Prairies. The leaves are eaten by white-tailed deer and cattle.

ASTERACEAE (Compositae)

TEXAS THISTLE
Cirsium texanum Buckl.

ANNUAL: From a taproot.
STEMS: With white, matted hairs.
LEAVES: Simple, alternate, with matted hairs below, dark green and glossy above, midvein area densely woolly; blades pinnatisect, uppermost leaves narrowly lanceolate; margins armed with sharp spines, lobe margins revolute.
INVOLUCRE: Rounded to cup-shaped; phyllaries armed with firm, recurved spines.
RECEPTACLE: With bristles; heads discoid.
RAY FLORETS: Absent.
DISK FLORETS: Corollas pink or white; pappus of bristles.
FRUIT: An achene 3.3–4.0 mm long.
COMMENTS: Abundant on a variety of soil types in the Rio Grande Plains and Coastal Prairies. The flower buds are occasionally eaten by white-tailed deer and the seeds are eaten by Rio Grande turkeys.

ASTERACEAE (Compositae)

Coreopsis L.

1a. Leaflets ovate, the terminal leaflet enlarged; leaves basal. *Coreopsis nuecensoides*

1b. Leaflets linear; stems leafy above. *Coreopsis tinctoria*

COREOPSIS, TICKSEED
Coreopsis nuecensoides Sm.

ANNUAL: From a taproot.

STEMS: Densely pubescent; leafless.

LEAVES: Simple, but odd-pinnately dissected, basal, pubescent; segments ovate, terminal leaflet enlarged; petioles densely pubescent.

INVOLUCRE: Phyllaries in 2 rows; outermost row narrowly triangular with translucent margins; the innermost row with phyllaries yellowish-green, subovate.

RECEPTACLE: Chaff present, bracts linear; deciduous; peduncle bearing a single, radiate head.

RAY FLORETS: Corollas yellow, variously lobed at the apex, with reddish-brown markings near the base; apparently sterile; ovary with 2 lateral wings.

DISK FLORETS: Corollas yellow; pappus of 2 awns.

FRUIT: An achene with lateral wings.

COMMENTS: Frequent on sandy loam and deep sandy soils in the eastern portion of the Rio Grande Plains and Coastal Prairies. The leaves are eaten by white-tailed deer and cattle.

ASTERACEAE (Compositae)

COREOPSIS, TICKSEED
Coreopsis tinctoria Nutt.
[Syn. *Coreopsis cardaminifolia* (DC.) T.&G.]

ANNUAL: From a taproot.
STEMS: Erect, freely branching above, glabrous.
LEAVES: Simple above, but usually appearing odd-pinnately compound below, opposite or in fascicles, glabrous; leaflets linear.
INVOLUCRE: Phyllaries in 2 rows; outermost row about 1/2 the length of the inner row, triangular, apex obtuse; innermost row united and of larger dimensions than the former; lobes brownish.
RECEPTACLE: Chaff linear, brownish; heads radiate, about 3 cm wide.
RAY FLORETS: Corollas yellow, with several lobes at the apex and with a red mark basally; sterile.
DISK FLORETS: Corollas reddish-purple to brown; pappus of 2 minute, lateral awns.
FRUIT: A flattened achene.
COMMENTS: Common in ditches, swales, and other low areas in the eastern portion of the Rio Grande Plains and Coastal Prairies. The leaves are occasionally eaten by white-tailed deer.

ENGELMANN DAISY
Engelmannia pinnatifida Nutt. ex Nutt.

PERENNIAL: From a woody rootstock.
STEMS: With septate, pustulate-based hairs.
LEAVES: Simple, alternate; hirsute with pustulate-based hairs; blades pinnatifid, midnerve white, conspicuous; lower leaves petioled, the uppermost clasping.
INVOLUCRE: Phyllaries in 2 rows; outermost row linear, pubescent; innermost row broad near the base but extended into a linear segment above.
RECEPTACLE: Chaff present, hirsute; heads radiate, 3.5–4.0 cm wide.
RAY FLORETS: Corollas yellow; 3-lobed at the apex, pistillate, ovary pubescent; pappus of 2 spines.
DISK FLORETS: Corollas yellow; perfect but not maturing achenes; pappus minute and scalelike; anthers dark brown.
FRUIT: An achene.
COMMENTS: Frequent on well drained sands and caliche soils in the Rio Grande Plains and Coastal Prairies. The foliage is eaten by cattle and white-tailed deer, and the seeds are eaten by several species of birds.

ASTERACEAE [Compositae]

CRUCITA, CHRISTMAS BUSH
Eupatorium odoratum L.

ANNUAL: Scandent; usually sprawling over other plants.
STEMS: Mostly glabrous, but occasionally with a few minute hairs.
LEAVES: Simple, alternate but usually with some opposite below, glabrous; blades lanceolate or oblanceolate; margins entire or with a few scattered teeth on lower leaves; petioles present.
INVOLUCRE: Cylindrical, about 1.0 CM long; lower phyllaries pubescent.
RECEPTACLE: Naked, convex; heads discoid, in corymbose clusters; peduncles pubescent.
RAY FLORETS: Absent.
DISK FLORETS: Corollas blue, fragrant; pappus of bristles.
FRUIT: A cylindrical achene.
COMMENTS: Frequent in brushy sites and sometimes openings on a variety of soil types in the eastern portion of the Rio Grande Plains and Coastal Prairies. The leaves are eaten by cattle and relished by white-tailed deer.

INDIAN BLANKET, FIREWHEEL
Gaillardia pulchella Foug.

ANNUAL: From a taproot, long-lived.
STEMS: Pubescent; shoots with a mild, resinous odor.
LEAVES: Simple, alternate, pubescent; blades lanceolate to oblanceolate, grayish-green; margins entire or slightly wavy; petioles present or absent.
INVOLUCRE: Phyllaries in 3 rows; pubescent, grayish-green; apex mucronate.
RECEPTACLE: Chaff acuminate at apex; heads radiate, 4–5 cm wide; peduncles pubescent.
RAY FLORETS: Corollas yellow above and red below, sometimes red almost to the apex; 3-lobed; pubescent on the back; sterile; pappus of spines, spines with lateral, translucent wings below.
DISK FLORETS: Corollas reddish-burgundy; anthers with yellow pollen.
FRUIT: An achene.
COMMENTS: Common on sandy soils in openings in the Rio Grande Plains and Coastal Prairies. The leaves are eaten by white-tailed deer.

ASTERACEAE (Compositae)

TEXAS BROOMWEED
Gutierrezia texana (DC.) T.&G.
[Syn. *Xanthocephalum texanum* (DC.) Shinners]

ANNUAL: From a taproot.
STEMS: Erect with numerous lateral branches above, ribbed, with scattered papillae, viscid.
LEAVES: Simple, alternate, glutinous; blades linear; margins entire; apex acute.
INVOLUCRE: Phyllaries lanceolate, green, glabrous.
RECEPTACLE: Chaff absent; heads radiate.
RAY FLORETS: Corollas yellow.
DISK FLORETS: Corollas yellow; pappus minute, scalelike.
FRUIT: A minutely pubescent achene.
COMMENTS: Abundant on various soils in a variety of habitats in the Rio Grande Plains and Coastal Prairies. The leaves are frequently eaten by white-tailed deer and cattle and the seeds are eaten by bobwhite quail. A noxious weed and aggressive invader on overgrazed ranges, especially following droughts.

ASTERACEAE (Compositae)

Helianthus L.
1a. Plants perennials, forming colonies from rhizomes; stems glabrous, glaucous. *Helianthus ciliaris*
1b. Plants annuals; stems pubescent. **2**

2a. Leaves mostly lanceolate to ovate-lanceolate. *Helianthus annuus*
2b. Leaves mostly triangular to ovate. *Helianthus praecox*

COMMON SUNFLOWER, MIRASOL
Helianthus annuus L.

ANNUAL: From a taproot; aromatic.
STEMS: Erect, branching at the nodes, hirsute, glutinous.
LEAVES: Simple, alternate, hirsute; blades lanceolate to ovate-lanceolate; petioles elongated, strigose.
INVOLUCRE: Phyllaries in 4–5 rows, lanceolate, pubescent; margins hirsute; apex acuminate.
RECEPTACLE: Convex; chaff nearly as long as disk floret; purple above with an acuminate apex; heads radiate.
RAY FLORETS: Corollas yellow, pistillate, sterile; ovary rudimentary; pappus of lateral awns.
DISK FLORETS: Corollas yellow below, with burgundy lobes, swollen at the base and constricted near the apex of the ovary; pubescent on the swollen segment; pappus of 2 lateral, deciduous, scalelike awns.
FRUIT: A minutely pubescent achene.
COMMENTS: Common on loam or clay soils in a variety of habitats in the Rio Grande Plains and Coastal Prairies. The seeds are eaten by bobwhite quail, mourning doves, Rio Grande turkeys, and white-winged doves, and the leaves are eaten by white-tailed deer and chachalacas.

ASTERACEAE (Compositae)

BLUEWEED SUNFLOWER
Helianthus ciliaris DC.

PERENNIAL: Often forming colonies from rhizomes.
STEMS: Erect, glabrous, glaucous, bluish-green.
LEAVES: Simple, opposite, sessile, glabrous above, but with scattered hairs below; blades linear to linear-lanceolate, glaucous; margins crisped with glandular hairs near the leaf base.
INVOLUCRE: Phyllaries in 3 rows, the outermost with recurved apices; margins ciliate.
RECEPTACLE: Chaff keeled, pubescent and bronze near the apex; heads radiate.
RAY FLORETS: Corollas yellow, sterile; pappus of several readily-dehiscent awns.
DISK FLORETS: Corollas reddish-bronze, pubescent below; style branches yellow; pappus similar to rays.
FRUIT: An achene.
COMMENTS: Occasional on clay soils in openings, fields, and waste places in the Rio Grande Plains and Coastal Prairies. The leaves are eaten by white-tailed deer.

SUNFLOWER, SAND SUNFLOWER
Helianthus praecox Engelm. & Gray ssp. *runyonii* (Heiser) Heiser
[Syn. *Helianthus debilis* Nutt. ssp. *runyonii* Heiser]

ANNUAL: From a taproot.
STEMS: Erect, hispid.
LEAVES: Simple, alternate, hispid; blades variously shaped but mostly triangular to ovate; margins toothed or lobed; petioles elongated.
INVOLUCRE: Phyllaries in 3 rows; outermost row acuminate or mucronate; innermost row abruptly acuminate, pubescent.
RECEPTACLE: Chaff present; heads radiate, up to 6 cm wide; peduncles hispid.
RAY FLORETS: Corollas yellow or yellow-orange; sterile; pappus of several awns; rudimentary ovary, angled.
DISK FLORETS: Corollas brown to bronze; pappus of 2 deciduous, scale-like awns; ovary pubescent.
FRUIT: An achene.
COMMENTS: Frequent on sandy soils in prairies and openings in the Rio Grande Plains and Coastal Prairies. The leaves are eaten by white-tailed deer and cattle, and the seeds are consumed by bobwhite quail, mourning doves, and Rio Grande turkeys.

ASTERACEAE (Compositae)

GRAY GOLDASTER
Heterotheca canescens (DC.) Shinners
[Syn. *Chrysopsis canescens* (DC.) T.&G.] [Syn. *Haplopappus canescens* DC.]

PERENNIAL: From a firm, woody base.
STEMS: Erect, densely canescent.
LEAVES: Simple, alternate or in fascicles; densely canescent; blades linear to lanceolate; margins entire; petioles indistinct.
INVOLUCRE: Phyllaries in 3–4 rows, linear to lanceolate, canescent.
RECEPTACLE: Slightly convex; heads radiate.
RAY FLORETS: Corollas yellow; pappus of bristles in 2 series; outer series small and inconspicuous; inner series longer, numerous, and with antrorse barbs; ovary densely ciliate.
DISK FLORETS: Corollas yellow; pappus and ovary similar to rays.
FRUIT: A densely ciliate achene.
COMMENTS: Frequent on sandy soils in prairies and openings in the northern Rio Grande Plains and Coastal Prairies. The leaves are eaten by white-tailed deer.

SEACOAST SUMPWEED, MARSHELDER
Iva annua L.

ANNUAL: From a taproot.
STEMS: Erect, scabrous with antrorsely oriented hairs; shoots aromatic.
LEAVES: Simple, alternate above, opposite below; scabrous; blades ovate to lanceolate, with 3 prominent veins arising from the petioles; margins toothed; apex acuminate.
INFLORESCENCE: Heads in spikes; discoid; bracts subtending heads lanceolate with ciliate margins; each head with several males and 1–2 females.
FRUIT: An achene.
COMMENTS: Frequent on various soils in low or damp areas in the Coastal Prairies. The young foliage is eaten by cattle.

ASTERACEAE [Compositae]

DEVIL-WEED, MEXICAN DEVIL-WEED, SPINY ASTER, WOLF WEED
Leucosyris spinosa (Benth.) Greene
[Syn. *Aster spinosus* Benth.]

PERENNIAL: From rhizomes, usually forming large colonies.
STEMS: Erect, glabrous, slightly angled.
LEAVES: (If present); simple, alternate, glabrous; blades awl-shaped.
INVOLUCRE: Phyllaries in 3–4 rows, triangular, glabrous.
RECEPTACLE: Chaff absent; heads radiate.
RAY FLORETS: Corollas white; pappus of numerous capillary bristles.
DISK FLORETS: Corollas yellow; pappus similar to rays.
FRUIT: An achene less than 1 mm long.
COMMENTS: Common on a variety of soil types in ditches, swales, lake margins, and other low areas in the Rio Grande Plains and Coastal Prairies. The young stems, leaves, and flowers are eaten by cattle and white-tailed deer.

GAYFEATHER, BLAZING STAR, BUTTON SNAKEROOT
Liatris mucronata DC.

PERENNIAL: From a woody base.
STEMS: Erect, glabrous, slightly glutinous.
LEAVES: Simple, alternate, glabrous; blades linear, punctate, glabrous, midvein conspicuous; margins entire; apex acute or acuminate.
INVOLUCRE: Cylindrical, 1.0–1.3 cm long; phyllaries ciliate on the margins; apex mucronate to acuminate.
RECEPTACLE: Chaff absent; heads discoid.
RAY FLORETS: Absent.
DISK FLORETS: Corollas with elongated lobes, purple; style branches longer than corolla lobes; pappus bristles numerous, 9–10 mm long; ovary pubescent.
FRUIT: A cylindrical, 4-angled achene.
COMMENTS: Infrequent on sandy loam or caliche soils often on slopes or bluffs in the northern portion of the Rio Grande Plains. The leaves are eaten by white-tailed deer.

ASTERACEAE (Compositae)

HOARY BLACKFOOT, BLACKFOOT DAISY
Melampodium cinereum DC.

PERENNIAL: From a woody base.
STEMS: Low-growing; densely pubescent with appressed hairs.
LEAVES: Simple, opposite, pubescent; blades narrowly lanceolate; margins somewhat revolute, broadly toothed or lobed; apex obtuse.
INVOLUCRE: Phyllaries 5, ovate, pubescent.
RECEPTACLE: Chaff yellow; heads radiate, about 1.5 cm wide.
RAY FLORETS: Corollas white, cleft at the apex, with greenish-yellow veins on the back, often lavender-tinged near the apex, pistillate and fertile; pappus bractlike, topped with a recurved spine.
DISK FLORETS: Corollas yellow, staminate; pollen yellow.
FRUIT: A tuberculate achene produced by ray florets.
COMMENTS: Frequent on sandy loam or caliche soils in the Rio Grande Plains. The leaves and flowers are an important food of white-tailed deer.

PALAFOXIA
Palafoxia texana DC. var. *ambigua* (Shinners) B.L. Turner

ANNUAL: From a taproot.
STEMS: Erect, branching freely; strigose.
LEAVES: Simple, alternate, pubescent; blades lanceolate; margins entire; apex acute or acuminate; petioles present.
INVOLUCRE: Phyllaries in 1–2 rows; strigose.
RECEPTACLE: Naked; heads discoid.
RAY FLORETS: Absent.
DISK FLORETS: Corollas lavender, tubular below and deeply lobed above; anthers black; style branches lavender; pappus of scales, 2–3 mm long.
FRUIT: A pubescent achene about 4 mm long.
COMMENTS: Frequent on sandy and sandy loam soils in the Rio Grande Plains and Coastal Prairies. The leaves are occasionally eaten by cattle and white-tailed deer.

ASTERACEAE (Compositae)

LYRELEAF PARTHENIUM, FALSE RAGWEED
Parthenium confertum Gray

PERENNIAL
STEMS: Woody below; densely pubescent with matted hairs.
LEAVES: Simple, with a basal cluster and alternate above, pubescent; blades lyrate; margins broadly lobed; petioles with an extension of blade margins.
INVOLUCRE: Phyllaries in 2–3 rows, densely pubescent.
RECEPTACLE: Naked; heads radiate, 5–7 mm wide.
RAY FLORETS: Corollas white, 0.3–0.5 mm long; pappus of 2 awns; ovary flattened, triangular.
DISK FLORETS: Corollas white, apparently infertile, pubescent.
FRUIT: A flattened achene.
COMMENTS: Common on sandy loam and caliche soils in the Rio Grande Plains and Coastal Prairies. The leaves are frequently eaten by white-tailed deer.

PAPERFLOWER, DUDWEED PAPERFLOWER
Psilostrophe gnaphalioides DC.

PERENNIAL
STEMS: White, woolly pubescent.
LEAVES: Simple, alternate, woolly pubescent; blades linear to lanceolate; margins entire or irregularly lobed below; apex obtuse.
INVOLUCRE: Phyllaries in one series, lanceolate, densely woolly pubescent.
RECEPTACLE: Chaff absent; heads radiate.
RAY FLORETS: Yellow; deeply 3-lobed; pistillate; ovary densely pubescent; pappus of several awns, each as long as the throat of the floret.
DISK FLORETS: Yellow; ovary and pappus similar to rays.
FRUIT: A densely woolly achene.
COMMENTS: Occasional on sandy loam or caliche soils in the Rio Grande Plains. The leaves and flowers are eaten by white-tailed deer. This herb is poisonous to sheep and cattle.

ASTERACEAE (Compositae)

MANYSTEM FALSE DANDELION
Pyrrhopappus multicaulis DC.
[Syn. *Pyrrhopappus geisieri* Shinners]

ANNUAL: From a taproot; all parts with a milky latex.
STEMS: With scattered, matted hairs.
LEAVES: Simple, basal, mostly glabrous, but with a few scattered hairs on the midvein; blades pinnatisect; margins with a few scattered hairs.
INVOLUCRE: Phyllaries in 3 rows, linear below and linear-lanceolate above, pubescent.
RECEPTACLE: Naked; heads ligulate.
RAY FLORETS: Corollas yellow, toothed at the apex; anthers brown; pappus of silky hairs.
DISK FLORETS: Absent.
FRUIT: An achene.
COMMENTS: Common on various soils in openings in the Rio Grande Plains and Coastal Prairies. The leaves and stems are eaten by cattle and white-tailed deer, and the seeds are eaten by Rio Grande turkeys.

UPRIGHT PRAIRIE CONEFLOWER, MEXICAN HAT
Ratibida columnifera (Nutt.) Woot. & Standl.
[Syn. *Ratibida columnaris* (Sims) D. Don., an illegitimate name]

PERENNIAL
STEMS: Erect, pubescent.
LEAVES: Simple, alternate, appressed pubescent; blades pinnatifid; segments lobed or unlobed.
INVOLUCRE: Phyllaries few, inconspicuous, lanceolate, pubescent; apex acute.
RECEPTACLE: Cone-shaped; chaff with paired, purple blotches; heads radiate; peduncle strongly angled, pubescent, about 15 cm long.
RAY FLORETS: Corollas few, up to 2 cm long, recurved toward the peduncle, in various shades of yellow and burgundy; male and female reproductive structures absent.
DISK FLORETS: Corollas about 0.5 mm long, with vertical purple lines; pappus of 2 lateral teeth.
FRUIT: A flattened achene.
COMMENTS: Common on clay, heavier loams, or caliche soils in the Rio Grande Plains and Coastal Prairies. The leaves are eaten by white-tailed deer and cattle, and the seeds are consumed by Rio Grande turkeys.

ASTERACEAE (Compositae)

BROWN-EYED SUSAN, CONEFLOWER
Rudbeckia hirta L.

ANNUAL: From a taproot.
STEMS: Erect, densely hirsute with papillose-based hairs.
LEAVES: Simple, alternate, densely hirsute; blades ovate to ovate-lanceolate; margins broadly toothed.
INVOLUCRE: Phyllaries in 1 row, hirsute; apex obtuse.
RECEPTACLE: Convex, hemispherical; chaff reddish; heads radiate; peduncle elongate, densely pubescent.
RAY FLORETS: Corollas yellow with a reddish zone near the base, lobed or unlobed at the apex, minutely pubescent on the back; sterile.
DISK FLORETS: Corollas reddish-brown; pappus absent.
FRUIT: A 4-angled achene.
COMMENTS: Occasional on deep sandy soils in the eastern Rio Grande Plains and Coastal Prairies. The leaves and stems are eaten by white-tailed deer and cattle, and the seeds are eaten by Rio Grande turkeys.

AWNLESS BUSH SUNFLOWER
Simsia calva (Engelm. & Gray) Gray

PERENNIAL: Slightly woody below.
STEMS: Erect, scabrous, with numerous short hairs and a few scattered longer hairs.
LEAVES: Simple, opposite, scabrous; blades triangular to ovate; margins broadly toothed; petioles scabrous.
INVOLUCRE: Phyllaries in 3–4 rows, the outermost recurved, lanceolate, with stiff, white hairs.
RECEPTACLE: Chaff nearly as long as the disk florets; heads radiate, 3-4 cm wide; peduncles elongated, bearing a single head.
RAY FLORETS: Corollas yellow, often burgundy on the lower surface, minutely pubescent; ovary awnless.
DISK FLORETS: Corollas yellow, pubescent; pappus of 2 stiff awns.
FRUIT: A columnar achene.
COMMENTS: Frequent on sandy loam and caliche soils in the Rio Grande Plains. The leaves are eaten by white-tailed deer.

ASTERACEAE (Compositae)

Sonchus L.
1a. Auricles on leaves rounded on the lower surface; achene surface not rough. *Sonchus asper*
1b. Auricles on leaves rounded above, but acute on the lower surface; achenes transversely rugulose. *Sonchus oleraceus*

PRICKLY SOWTHISTLE, ACHICORIA DULCE
Sonchus asper (L.) Hill.

ANNUAL: From a taproot; all parts with a milky latex.
STEMS: Erect, glabrous.
LEAVES: Simple, alternate, glabrous; blades pinnatisect or pinnatifid; margins spiny; petioled leaves with lateral wings, uppermost leaves subperfoliate, with rounded, remotely toothed auricles.
INVOLUCRE: Phyllaries in several rows, lanceolate.
RECEPTACLE: Chaff absent; heads ligulate.
RAY FLORETS: Corollas yellow, fringed at the apex, the tube pubescent; pappus of white, cottony hairs.
DISK FLORETS: Absent.
FRUIT: A brown, ribbed achene with lateral wings; surface not rugulose.
COMMENTS: Frequent on various soils in openings and disturbed areas in the Rio Grande Plains and Coastal Prairies. The leaves are occasionally eaten by white-tailed deer and Rio Grande turkeys.

COMMON SOWTHISTLE
Sonchus oleraceus L.

Similar to *S. asper;* however, it has leaf auricles that are rounded above, but acute at the base. The achenes are transversely rugulose. This species occurs in the same habitat as the above species and is also eaten by white-tailed deer. [Not illustrated.]

ASTERACEAE (Compositae)

SHOWY NERVE-RAY, SQUARE-BUD DAISY
Tetragonotheca repanda (Buckl.) Small

PERENNIAL: From a woody rootstock.
STEMS: Erect, minutely pubescent.
LEAVES: Simple, basal, opposite above, glabrous above and pubescent below; blades ovate; margins toothed; basal leaves with long petioles; uppermost leaves sessile.
INVOLUCRE: Phyllaries 4 (resembles a large bud).
RECEPTACLE: Convex; chaff yellow-green; heads radiate.
RAY FLORETS: Corollas yellow, pistillate and fertile; pappus a scalelike ring.
DISK FLORETS: Corollas yellow; pappus similar to rays.
FRUIT: A cylindrical, pubescent achene.
COMMENTS: Common on deep sands in the Rio Grande Plains and Coastal Prairies. The leaves and flower buds are occasionally eaten by white-tailed deer.

GREEN THREAD
Thelesperma nuecense B.L. Turner

ANNUAL: From a taproot.
STEMS: Erect, glabrous.
LEAVES: Compound, opposite, glabrous; leaflets linear, threadlike, about 1 mm wide.
INVOLUCRE: Phyllaries in 2 rows; outermost row linear, green-glaucous; innermost row broadly ovate, veins purple, margins translucent.
RECEPTACLE: Chaff translucent with 2 parallel, bronze veins; heads radiate; peduncles elongated.
RAY FLORETS: Corollas yellow, with a bright red spot near the base; 3-lobed at the apex; sterile; ovary pubescent on margins.
DISK FLORETS: Corollas reddish-bronze; pappus of 2 barbed awns.
FRUIT: An achene with a papillose surface.
COMMENTS: Common on sandy soils in the eastern Rio Grande Plains and Coastal Prairies. The leaves and flower buds are eaten by white-tailed deer.

ASTERACEAE (Compositae)

BRISTLELEAF DOGWEED
Thymophylla tenuiloba (DC.) Small
[Syn. **Dyssodia tenuiloba** (DC.) Robins.]

ANNUAL or PERENNIAL: With a strong, marigold-like odor.
STEMS: Low-growing; usually branching at the nodes; pubescent.
LEAVES: Compound, alternate, pubescent; blades dissected into linear segments 1 mm or less wide; sessile.
INVOLUCRE: Inflorescence axis elongate with scattered, appressed, narrowly lanceolate, pubescent bracts; phyllaries in 2 series; the lowermost free, lanceolate, margins ciliate, apex acuminate; larger phyllaries united into a cuplike structure; apex with equilaterally-triangular lobes; lobes minutely pubescent, margins scarious, with paired, linear glands.
RECEPTACLE: Naked; heads radiate.
RAY FLORETS: Corollas yellow, slightly notched at the apex; pappus of bristles.
DISK FLORETS: Corollas yellow, pappus similar to rays.
FRUIT: An achene, truncate at the apex, pointed at the base.
COMMENTS: Frequent on sandy loam and caliche soils in the Rio Grande Plains and Coastal Prairies. The leaves are occasionally eaten by white-tailed deer.

Thymophylla tenuiloba (DC.) Small var. *treculi* (Gray) Strother [Syn. *Dyssodia tenuiloba* (DC.) Small var. *wrightii* (Gray) Strother; *Dyssodia wrightii* (Gray) B. Robins.] This variety has a pappus of 5 short and 5 long bristles. [Not illustrated.]

ASTERACEAE (Compositae)

Verbesina L.
1a. Corollas of ray and disk florets white; stems with lateral wings.
   Verbesina microptera
1b. Corollas of ray and disk florets yellow; stems lacking wings. **Verbesina encelioides**

COWPEN DAISY, GOLDEN CROWNBEARD
Verbesina encelioides (Cav.) Benth. & Hook. ex Gray
[Syn. **Ximensia encelioides** Cav.]

ANNUAL: From a taproot.
STEMS: New growth strigose-pubescent.
LEAVES: Simple, alternate, grayish-green, pubescent; blades lanceolate or ovate; margins dentate-serrate; base cordate or truncate; petioles present.
INVOLUCRE: Phyllaries linear, pubescent; the larger nearly as long as the rays.
RECEPTACLE: Slightly convex; chaff present; heads radiate.
RAY FLORETS: Corollas yellow, 3-lobed at the apex, throat and tube pubescent, pistillate.
DISK FLORETS: Corollas yellow, tube pubescent below; pappus of 2 bristles.
FRUIT: A flattened, pubescent achene.
COMMENTS: Common on sandy loam and sandy soils in prairies, openings and disturbed areas in the Rio Grande Plains and Coastal Prairies. The seeds are eaten by bobwhite quail and Rio Grande turkeys.

FROSTWEED, CAPITANA
Verbesina microptera DC.

PERENNIAL: From short rhizomes.
STEMS: Erect, hirsute, with broad lateral wings.
LEAVES: Simple, alternate, hirsute; blades oblanceolate to ovate; margins crenate; petioles extended as wings along the stems.
INVOLUCRE: Phyllaries in 2–3 rows; linear; pubescent.
RECEPTACLE: Convex; chaff present; heads radiate.
RAY FLORETS: Corollas white; 3-lobed at the apex; tube pubescent below; pistillate; pappus of 2 awns.
DISK FLORETS: Corollas white or greenish-white; pubescent below; pappus of 2 awns.
FRUIT: An achene.
COMMENTS: Common on various soils in openings and brushy sites in the eastern portion of the Rio Grande Plains and Coastal Prairies. The leaves and stems are eaten by white-tailed deer and cattle.

ASTERACEAE (Compositae)

ORANGE ZEXMENIA
Wedelia hispida H.B.K.
[Syn. *Zexmenia hispida* (H.B.K.) Gray]

PERENNIAL
STEMS: Erect, scabrous.
LEAVES: Simple, opposite, scabrous or strigose with pustulate-based hairs; blades lanceolate; apex acute or acuminate; leaves turning black after drying.
INVOLUCRE: Phyllaries strigose.
RECEPTACLE: Chaff keel-shaped, midvein reddish-purple; heads radiate.
RAY FLORETS: Corollas yellow or orange; pappus spinose, ciliate on the spine margins.
DISK FLORETS: Corollas yellow to yellow-orange; pappus of 2 spines about 1/2 –2/3 length of disk florets.
FRUIT: A ciliate, pubescent achene.
COMMENTS: Frequent on various soils in openings and brushy sites in the Rio Grande Plains and Coastal Prairies. The leaves are eaten by cattle and white-tailed deer.

TEXAS SLEEPY-DAISY
Xanthisma texanum DC.

ANNUAL: From a taproot.
STEMS: Minutely pubescent.
LEAVES: Simple, alternate, glabrous, appressed against the stem, smaller near the stem apex; blades lanceolate; margins ciliate.
INVOLUCRE: Phyllaries in several rows, acuminate or mucronate at apex, often recurved.
RECEPTACLE: Chaff bristly; heads radiate.
RAY FLORETS: Corollas yellow; pappus of narrow scales and bristles.
DISK FLORETS: Corollas yellow; pappus similar to rays.
FRUIT: An achene.
COMMENTS: Frequent on loam and sandy soils in the Rio Grande Plains and Coastal Prairies. The leaves and flowers are eaten by white-tailed deer.

BRASSICACEAE (Cruciferae)

1a. Capsules linear. **2**
1b. Capsules triangular, globose, or round and flattened. **3**

2a. Shoots densely pubescent with stellate and/or glandular-tipped hairs; leaf segments about 1–2 mm wide; capsules about 5 mm long. *Descurania*
2b. Shoots glabrous or nearly so; leaf segments much wider than those of the above; capsules 10–20 mm long. *Rorippa*

3a. Capsules inflated, globose. *Lesquerella*
3b. Capsules flattened. **4**

4a. Capsule and ovary densely pubescent; not notched at the apex. *Synthlipsis*
4b. Capsule and ovary glabrous; can be notched at the apex. **5**

5a. Capsule widest near the apex, tapered at the base, roughly triangular; corolla well-developed. *Capsella*
5b. Capsule rounded; corolla absent or minute. *Lepidium*

SHEPHERD'S PURSE
Capsella bursa-pastoris (L.) Medic.

ANNUAL: From a taproot.
STEMS: Erect, glabrous, but with a few, scattered, stellate hairs.
LEAVES: Simple, alternate; blades oblanceolate, pinnatisect below; margins irregularly-toothed; lowermost leaves petioled, uppermost sessile, paired auricles present; auricles acute at apex.
INFLORESCENCE: A raceme with glabrous pedicels; axis mostly glabrous.
  CALYX: Sepals 4, with a few scattered hairs; apex pink or purple-tinged, obtuse.
  COROLLA: Petals 4, free, white, slightly clawed below, apex obtuse.
  STAMENS: 6, 4 long and 2 short; anthers greenish-yellow.
  PISTIL: Ovary superior; style 1, unbranched.
FRUIT: A flattened capsule, apex notched, tapered below to the pedicel; many-seeded, septum white, persistent.
COMMENTS: Locally abundant in waste places and old fields in the eastern Rio Grande Plains and Coastal Prairies. The leaves and flowers are eaten by white-tailed deer, and the leaves and seeds are consumed by Rio Grande turkeys.

BRASSICACEAE (Cruciferae)

PINNATE TANSY MUSTARD
Descurainia pinnata (Walt.) Britt.

ANNUAL: From an elongated taproot.
STEM: Erect, with branched, glandular-tipped hairs.
LEAVES: Bipinnately compound, alternate, stellate-pubescent; leaflet segments 1–2 mm wide.
INFLORESCENCE: A raceme; pedicels elongated, glabrous.
    CALYX: Sepals 4, yellowish-green; apex rounded.
    COROLLA: Petals 4, free, yellowish-green; apex obtuse.
    STAMENS: 6, not all of the same length.
    PISTIL: Ovary superior; style 1, unbranched; stigma capitate.
FRUIT: An elongated capsule with a central, longitudinal septum.
COMMENTS: Frequent on sandy soils in the Rio Grande Plains and Coastal Prairies. The leaves and flowers are eaten by white-tailed deer, and the seeds are occasionally eaten by bobwhite quail and Rio Grande turkeys.

PEPPERWEED, LENTEJILLA
Lepidium virginicum L. var. *medium* (Greene) C.L. Hitchc.
[Syn. *Lepidium medium* Greene]

ANNUAL: From a taproot.
STEMS: Erect, glabrous.
LEAVES: Simple, alternate, mostly glabrous; blades pinnatisect below, linear to narrowly lanceolate above; margins of uppermost leaves remotely toothed or entire.
INFLORESCENCE: A raceme.
    CALYX: Sepals 4, free, pubescent, less than 1 mm long; margins white.
    COROLLA: Petals absent or only rudimentary.
    STAMENS: 2, free.
    PISTIL: Ovary superior; style 1, unbranched.
FRUIT: A rounded, flattened capsule.
COMMENTS: Common on various soils in prairies, openings, and waste places in the Rio Grande Plains and Coastal Prairies. The leaves and fruits are eaten by cattle and white-tailed deer, whereas the seeds are eaten by several species of birds.

BRASSICACEAE (Cruciferae)

Lesquerella S. Wats
1a. Leaves silvery-gray; plants woody near the base; fruiting pedicles
    S-shaped. ***Lesquerella argyraea***
1b. Plants not as above. **2**

2a. Pedicels recurving near the apex in fruiting racemes; mature fruits
    pointing downward; capsules nearly flattened when mature. ***Lesquerella***
    lasiocarpa
2b. Pedicels not as above; fruits usually not pointed downward at maturity;
    capsules globose at maturity. ***Lesquerella lindheimeri***

SILVER BLADDERPOD
Lesquerella argyraea (Gray) Wats.

PERENNIAL: From a woody rootstock.
STEMS: Low-growing; silvery-gray with closely-appressed stellate hairs.
LEAVES: Simple, alternate, with pubescence similar to stems; blades
lanceolate, 1–2 cm long; margins entire; petioles present.
INFLORESCENCE: A raceme with pubescent pedicels.
    CALYX: Sepals 4, free, stellate pubescent.
    COROLLA: Petals 4, free, yellow, actinomorphic.
    STAMENS: 6; 4 long and 2 short.
    PISTIL: Ovary superior; style 1, unbranched.
FRUIT: An inflated, rounded capsule; fruiting pedicels S-shaped; style
persistent.
COMMENTS: Found on sandy and calcareous soils in the Rio Grande
Plains. The leaves are eaten by white-tailed deer, and the seeds are
consumed by scaled quail.

BRASSICACEAE (Cruciferae)

ROUGHPOD BLADDERPOD
Lesquerella lasiocarpa (Gray) S. Wats.

ANNUAL: Reported also as a biennial or perennial.
STEMS: Erect or reclining; pubescent with a mixture of branched and unbranched hairs.
LEAVES: Simple, alternate, pubescent; blades usually broadly lanceolate; margins lobed; sessile or only occasionally petioled.
INFLORESCENCE: An elongated raceme; pedicels recurving near the apex in fruiting racemes; mature fruits pointing downward.
    CALYX: Sepals 4, free; pubescent; margins translucent; apex obtuse.
    COROLLA: Petals 4, free, yellow, broadly-rounded, slightly clawed at base; lateral appendages present.
    STAMENS: 6, about equal in length.
    PISTIL: Ovary superior, style 1, unbranched; stigma capitate.
FRUIT: A nearly flattened capsule; pubescent, with larger, unbranched hairs and smaller stellate hairs.
COMMENTS: Frequent on sandy, gravelly, and/or disturbed soils in the Rio Grande Plains. The leaves, flowers, and fruits are eaten by white-tailed deer, and the seeds are eaten by bobwhite quail.

LINDHEIMER BLADDERPOD
Lesquerella lindheimeri (Gray) S. Wats.

Similar to *L. lasiocarpa*; however, the fruits are globose at maturity and the pedicels bearing the mature fruits are usually recurved. The species is common on clay and loamy soils in the Coastal Prairies and eastern portion the Rio Grande Plains. The leaves are eaten by white-tailed deer. [Not illlustrated.]

TANSYLEAF YELLOWCRESS
Rorippa teres (Michx.) Stuckey

ANNUAL: From a taproot.
STEMS: Low-growing; with only a few scattered hairs.
LEAVES: Simple, alternate, pubescent; blades pinnatifid or pinnatisect; lobes toothed.
INFLORESCENCE: A raceme with glabrous axis and pedicels.
    CALYX: Sepals 4, free.
    COROLLA: Petals 4, free, yellow, to 2 mm long.
    STAMENS: 6, 4 long and 2 short.
    PISTIL: Ovary superior; style 1, unbranched.
FRUIT: A linear capsule with a persistent style.
COMMENTS: Frequent on sandy soils along lake margins, stream banks, swales, and other low places in the Coastal Prairies and Rio Grande Plains. The leaves and flowers are occasionally eaten by white-tailed deer.

BRASSICACEAE (Cruciferae)

GREGG KEELPOD
Synthlipsis greggii Gray

ANNUAL: From a taproot.
STEMS: Sprawling; stellate-pubescent.
LEAVES: Simple, alternate, stellate-pubescent; blades mostly ovate; margins irregularly lobed; petioles present.
INFLORESCENCE: A few-flowered raceme.
    CALYX: Sepals 4, free, stellate-pubescent.
    COROLLA: Petals 4, free, white turning rose or pink, about 1 cm long.
    STAMENS: 6; anthers yellow.
    PISTIL: Ovary superior, densely pubescent; style 1, unbranched.
FRUIT: A densely stellate-pubescent capsule, about 1 cm long and 5 mm wide.
COMMENTS: Infrequent on sandy or gravelly soils in the western half of the Rio Grande Plains. White-tailed deer occasionally eat the leaves and flowers.

CAMPANULACEAE

Triodanis Raf.
1a. Capsule dehiscing from a linear slit on the side of the capsule. *Triodanis holzingeri*
1b. Capsule dehiscing from a pore near the apex of the capsule. *Triodanis perfoliata*

VENUS' LOOKING-GLASS
Triodanis holzingeri McVaugh
[Syn. *Specularia holzingeri* (McVaugh) Fern.]

ANNUAL: From a taproot.
STEMS: Erect, ribbed, with stiff hairs arising at right angles to the axis.
LEAVES: Simple, alternate, pubescent to nearly glabrous; blades ovate; margins toothed; apex nearly obtuse.
INFLORESCENCE: Flowers solitary in leaf axils.
    CALYX: Sepals 5, united.
    COROLLA: Petals 5, united, purple-violet.
    STAMENS: 5, epipetalous.
    PISTIL: Ovary inferior; style 1, branches 3.
    FRUIT: A dehiscent capsule, dehiscing from a linear slit on side of capsule.
COMMENTS: Frequent on sand or loamy soils in openings, waste places, and along roads in the Rio Grande Plains and western portion of the Coastal Prairies. The leaves are eaten by white-tailed deer.

CAMPANULACEAE

CLASPING VENUS' LOOKING-GLASS
Triodanis perfoliata (L.) Nieuw. var. *biflora* (R.&P.) Greene
[Syn. *Specularia biflora* (L.) DC.]

ANNUAL: From a taproot.
STEMS: Erect, with minute, translucent wing; minutely, retrorsely scabrous.
LEAVES: Simple, alternate, glabrous; blades ovate; margins toothed; apex obtuse.
INFLORESCENCE: Axillary, 2–3 flowers per node, sessile or nearly so; linear; green bracts subtend calyx.
CALYX: Sepals 3, 4, 5; united; lobes retrorsely scabrous, lanceolate; apex acute.
COROLLA: Petals 5, occasionally 6, blue, actinomorphic.
STAMENS: 5, epipetalous.
PISTIL: Ovary inferior; style 1, branches 3.
FRUIT: A many-seeded capsule, dehiscent by a pore near apex; sepals persistent on capsule.
COMMENTS: Frequent on sandy soils in prairies, openings, and along roads in the Coastal Prairies. The leaves are consumed by white-tailed deer.

CARYOPHYLLACEAE

PROSTRATE STARWORT, LLOVISNA
Stellaria prostrata Baldw.

ANNUAL: From a taproot.
STEMS: Low-growing, creeping and sprawling; glabrous or pubescent.
LEAVES: Simple, opposite, mostly glabrous; blades ovate or cordate; margins entire; apex acute; petioles with a few, scattered glandular hairs.
INFLORESCENCE: Flowers 1-2 in leaf axils; pedicels glandular-pubescent.
CALYX: Sepals 5, free, glandular pubescent, margins entire.
COROLLA: Petals 5, free, each divided nearly 3/4 its length, white.
STAMENS: 10; anthers white.
PISTIL: Ovary superior; styles 3.
FRUIT: A capsule; seeds round with a barbed surface.
COMMENTS: Frequent on various soils in pastures and woods in the Coastal Prairies and eastern portion of Rio Grande Plains, usually shaded. The leaves are occasionally eaten by white-tailed deer, and the seeds are known to be eaten by scaled quail.

COMMELINACEAE

1a. Inflorescence arising from a folded green bract; petals unequal.
   Commelina
1b. Inflorescence not as above; petals equal. **2**

2a. Stems trailing and rooting at the nodes; shoots succulent; inflorescence
   a cyme arising from the stem apex. **Callisia**
2b. Stems absent; not succulent; inflorescence an umbellate cyme arising
   from the upper leaf axils. **Tradescantia**

LITTLE-FLOWER SPIDERWORT
Callisia micrantha (Torr.) Hunt
[Syn. *Tradescantia micrantha* Torr.]

PERENNIAL: From fibrous roots.
STEMS: Trailing and rooting at the nodes; succulent; glabrous to minutely
pubescent.
LEAVES: Simple, alternate; blades lanceolate, sessile, venation parallel;
margins pubescent with longer hairs near the sheath; sheath pubescent.
INFLORESCENCE: Few-flowered cymes arising from stem apex; pedicles
subtended by several linear bracts.
   CALYX: Sepals 3, free, keel-shaped; glabrous but hirsute on the midvein.
   COROLLA: Petals 3, free; rose, pink or rose-purple; actinomorphic.
   STAMENS: 6, free; filaments with purple hairs; anthers yellow.
   PISTIL: Ovary superior; style 1, unbranched; stigma capitate.
FRUIT: A 3-lobed capsule.
COMMENTS: Frequent on various soils in the Coastal Prairies and extreme
eastern portion of the Rio Grande Plains. The leaves and stems are eaten by
white-tailed deer, javelinas, feral pigs, and sandhill cranes.

COMMELINACEAE

Commelina L.
1a. Leaves linear to lanceolate. ***Commelina erecta***
1b. Leaves ovate to broadly ovate-lanceolate. ***Commelina virginica***

NARROWLEAF DAYFLOWER, WIDOW'S TEARS
Commelina erecta L. var. *angustifolia* (Michx.) Fern.

PERENNIAL: From fleshy, fibrous roots.
STEMS: Erect, but falling in larger plants; glabrous.
LEAVES: Simple, alternate, sheathing the upper stems; blades linear to lanceolate, venation parallel; sheaths pubescent but glabrous near the auricle.
INFLORESCENCE: 2–3 flowers arising from a folded, green bract; bract laterally acute, rounded at the apex.
    CALYX: Sepals 3, free, unequal and nearly transparent.
    COROLLA: Petals 3, free, unequal, blue or blue and white, fragile; wilting into the bracts in the afternoon.
    STAMENS: 6, free; anthers bright yellow.
    PISTIL: Ovary superior; style 1, unbranched.
FRUIT: A capsule.
COMMENTS: Common on a variety of soil types in prairies, openings, stream bottoms, and along roads throughout the Coastal Prairies and Rio Grande Plains. The leaves and stems are an important food of white-tailed deer and the seeds are eaten by bobwhite quail, white-winged doves, and mourning doves. Narrowleaf dayflower is also eaten by cattle.

TROPICAL DAYFLOWER, VIRGINIA DAYFLOWER
Commelina virginica L.
[Syn. *Commelina elegans* H.B.K.]

PERENNIAL
STEMS: Erect, but larger plants reclining and often rooting at the nodes; glabrous.
LEAVES: Simple, alternate, glabrous; blades ovate to broadly ovate-lanceolate; margins entire; apex acute or obtuse; petioles present.
INFLORESCENCE: Several flowers arising from a folded bract.
    CALYX: Sepals 3, free.
    COROLLA: Petals 3, light blue-violet; quickly deteriorating.
    STAMENS: 3; anthers orange.
    PISTIL: Ovary superior.
FRUIT: A capsule.
COMMENTS: Frequent on loamy soils or caliche in pastures and woods in the western portion of the Coastal Prairies and southern and eastern portions of the Rio Grande Plains. The leaves and stems are eaten by white-tailed deer and cattle.

COMMELINACEAE

STEMLESS SPIDERWORT
Tradescantia subacaulis Bush
[Syn. *Tradescantia texana* Bush]

PERENNIAL
STEMS: Absent.
LEAVES: Simple, basal or alternate; shaggy-pubescent; blades linear (grasslike), venation parallel.
INFLORESCENCE: An umbellate cyme, arising from the upper leaf axils; pedicels pubescent, elongated.
    CALYX: Sepals 3, free, pubescent.
    COROLLA: Petals 3, free, blue, 1.0–1.5 cm long, wilting quickly.
    STAMENS: Up to 12, free; filaments pubescent.
    PISTIL: Ovary superior; style 1, unbranched.
FRUIT: A capsule.
COMMENTS: Localized on sandy soils in the Coastal Prairies and extreme eastern portion of the Rio Grande Plains. The leaves and stems are eaten by white-tailed deer.

CONVOLVULACEAE

1a. Styles 2, each with 2 branches; corolla without a prominent tube.
Evolvulus
1b. Style 1; corolla with a prominent tube. **2**

2a. Style unbranched; stems with irregularly spaced, pustulate-based hairs.
Ipomoea
2b. Style with 2 branches; stems with matted white hairs. *Convolvulus*

TEXAS BINDWEED, GRAY BINDWEED
Convolvulus equitans Benth.
[Syn. *Convolvulus hermannioides* Gray]

PERENNIAL
STEMS: Prostrate, trailing and twining; tendrils absent; grayish-green; pubescent with matted, white hairs.
LEAVES: Simple, alternate, densely pubescent; blades variously shaped but usually lanceolate to ovate, grayish-green; margins lobed; bases often auriculate; petioles densely pubescent.
INFLORESCENCE: Flowers usually solitary at the nodes; pedicle pubescent, elongated.
    CALYX: Sepals 5, united near the base, imbricate, pubescent; apex rounded or obtuse.
    COROLLA: Petals 5, united, actinomorphic, tubular, white or pinkish with red or white in the throat.
    STAMENS: 5, epipetalous, slightly unequal; anthers brown.
    PISTIL: Ovary superior; style 1, branches 2.
FRUIT: A capsule.
COMMENTS: Frequent on sandy soils or caliche in openings, prairies, and disturbed places throughout the Coastal Prairies and Rio Grande Plains. The leaves are eaten by white-tailed deer, and the seeds are consumed by bobwhite quail.

CONVOLVULACEAE

Evolvulus L.
1a. Corollas blue. ***Evolvulus alsinoides***
1b. Corollas white. ***Evolvulus sericeus***

SLENDER EVOLVULUS
Evolvulus alsinoides (L.) L. var. *hirticaulis* Torr.

PERENNIAL
STEMS: Trailing, but not rooting at the nodes; silky pubescent.
LEAVES: Simple, alternate, pubescent, sessile; blades ovate; margins entire.
INFLORESCENCE: Flowers solitary from the leaf axils; pedicels pubescent.
    CALYX: Sepals 5, united, pubescent.
    COROLLA: Petals 5, united, light to dark blue.
    STAMENS: 5, epipetalous, white.
    PISTIL: Ovary superior; styles 2, branches 2 per style.
FRUIT: A rounded capsule with 4 carpels and 4 seeds.
COMMENTS: Frequent on sandy soils or caliche in the western portion of the Coastal Prairies and Rio Grande Plains. The leaves are eaten by white-tailed deer and the seeds are consumed by bobwhite quail.

SILKY EVOLVULUS
Evolvulus sericeus Sw.

PERENNIAL
STEMS: Trailing, not rooting at the nodes; silky pubescent.
LEAVES: Simple, alternate, subsessile, silky pubescent below, glabrous above; blades lanceolate.
INFLORESCENCE: Flowers usually solitary at the nodes; pedicel silky pubescent.
    CALYX: Sepals 5, united below, persistent in fruit; pubescent.
    COROLLA: Petals 5, united into a tube, white or light bluish-white, up to 1.5 cm in diameter.
    STAMENS: 5, epipetalous, white.
    PISTIL: Ovary superior; styles 2, each with 2 branches.
FRUIT: A rounded capsule, with 4 carpels and 4 seeds.
COMMENTS: Frequent on sandy soils or caliche in the Rio Grande Plains. The seeds are eaten by bobwhite quail and the foliage is occasionally consumed by white-tailed deer.

CONVOLVULACEAE

SHARPPOD MORNING GLORY
Ipomoea trichocarpa Ell.

PERENNIAL

STEMS: Trailing, twining, climbing; with irregularly spaced, pustulate-based hairs.

LEAVES: Simple, alternate, with pubescence similar to that of stems; blades usually 3-lobed, base cordate or auriculate; petioles elongated.

INFLORESCENCE: Flowers solitary from the leaf axils; pedicels elongated.

   CALYX: Subtended by several linear bracts; sepals 5, united near the base; margins ciliate; apex subulate.

   COROLLA: Petals 5, united near the base, trumpet-shaped, rose to reddish-purple, dark red in the throat.

   STAMENS: 5, epipetalous, with tufted hairs near the base of the filaments.

   PISTIL: Ovary superior; style 1, unbranched.

FRUIT: A capsule.

COMMENTS: Common on a variety of soil types in pastures, woods, stream bottoms and waste places throughout the Rio Grande Plains and Coastal Prairies. The seeds are eaten by bobwhite quail and the leaves are consumed by white-tailed deer.

CUCURBITACEAE

GLOBE BERRY, DEER APPLES
Ibervillea lindheimeri (Gray) Greene
[Syn. *Sicydium lindheimeri* Gray]

PERENNIAL: A vine, from an enlarged taproot in older plants.
STEMS: Slightly angled, glabrous, dark green.
LEAVES: Tendrils present; leaves alternate, glabrous above and pustulate-glandular below; blades deeply palmately dissected, 3–5 lobed, each lobe, in turn, lobed; margins with widely spaced, whitish papillae.
INFLORESCENCE: Male flowers solitary or in racemose clusters.
STAMINATE FLOWERS: Plants dioecious.
    CALYX: Sepals 5, united.
    COROLLA: Petals 5, united, yellow-green.
    STAMENS: 3, epipetalous.
PISTILLATE FLOWERS: Female flowers solitary.
    CALYX: Sepals similar to the staminate flowers.
    COROLLA: Petals similar to the staminate flowers; pubescent on the inner surfaces.
    PISTIL: Ovary inferior; stigma 3-lobed.
FRUIT: A red or orange, malodorous berry; about 2.5 cm in diameter; pulp yellowish.
COMMENTS: Frequent on a variety of soil types in woodlands or thickets in the Rio Grande Plains and western portion of Coastal Prairies. The seeds are eaten by scaled quail, and the leaves are occasionally eaten by white-tailed deer.

EUPHORBIACEAE

1a. Plants with irritating hairs. **2**
1b. Plants not as above. **3**

2a. Plant a vine; calyx green; milky latex absent. *Tragia*
2b. Plant erect, not vinelike; calyx white; milky latex present. *Cnidoscolus*

3a. Plants dioecious. **4**
3b. Plants monoecious. **5**

4a. Calyx of male flowers and styles of female flowers red; leaf margins
   toothed. *Acalypha*
4b. Calyx and styles not as above; leaf margins entire. *Croton*

5a. Shoots pubescent but lacking irritating hairs. **6**
5b. Shoots glabrous. **7**

6a. Shoots stellate pubescent; plants annuals. *Croton*
6b. Shoots with malpighiaceous hairs (hairs T-shaped); plants perennials.
   Argythamnia

7a. Leaf blades deeply palmately lobed or dissected; inflorescence a cyme;
   corolla red; shoots arising from an enlarged tuberlike root. *Jatropha*
7b. Leaf blades entire or toothed, not deeply dissected; inflorescence not as
   above; corolla absent; annuals or perennials. **8**

8a. Calyx absent; inflorescence a petaloid cyathium. *Euphorbia*
8b. Calyx present; inflorescence a spike or flowers 1–2 from leaf axils. **9**

9a. Inflorescence a spike; leaf margins toothed; milky latex present.
   Stillingia
9b. Inflorescence with 1–2 flowers from leaf axils; leaf margins entire; milky
   latex absent. *Phyllanthus*

EUPHORBIACEAE

ROUND COPPERLEAF, YERBA DE LA RABIA, CARDINAL FEATHER
Acalypha radians Torr.

PERENNIAL: Dioecious.
STEMS: Low-growing; densely pubescent.
LEAVES: Simple, alternate, pubescent; blades with palmate venation; margins toothed; truncate at the base.
INFLORESCENCE: Males in a terminal spike; females in terminal glomerules.
    CALYX: Males, sepals 4, minute, pubescent, red; females, 3, minute, pubescent, green.
    COROLLA: Lacking in both sexes.
    STAMENS: Numerous.
    PISTIL: Ovary superior; styles 3, branched, red.
FRUIT: A 3-seeded capsule.
COMMENTS: Locally abundant on sandy soils in openings and prairies in the Rio Grande Plains and Coastal Prairies. The seeds are eaten by bobwhite quail and Rio Grande turkeys, and the leaves are eaten by white-tailed deer.

EUPHORBIACEAE

WILD MERCURY
Argythamnia humilis (Engelm. & Gray) Muell. Arg.

PERENNIAL: With a woody base; monoecious.
STEMS: Erect, sprawling or reclining; with long malpighiaceous, white hairs.
LEAVES: Simple, alternate, with pubescence similar to stems; blades lanceolate; apex obtuse; petioles present.
INFLORESCENCE: In few-flowered axillary racemes; females below the males.
STAMINATE FLOWERS
    CALYX: Sepals 5, free, ciliate, to 2 mm long.
    COROLLA: Petals absent.
    STAMENS: 8–9, in 2 series; anthers yellow.
PISTILLATE FLOWERS
    CALYX: Sepals 5, free, densely ciliate.
    COROLLA: Petals absent.
    PISTIL: Ovary superior, 3-lobed, densely pubescent; styles 3, each with 2 minute branches.
FRUIT: A 3-lobed capsule; each carpel with 1 seed.
COMMENTS: Common on sandy soils in openings, prairies, and waste places in the Rio Grande Plains and Coastal Prairies. The seeds are an important food of bobwhite quail and mourning doves, and the leaves are eaten by white-tailed deer and cattle.

EUPHORBIACEAE

TEXAS BULLNETTLE, MALA MUJER
Cnidoscolus texanus (Muell. Arg.) Small

PERENNIAL: From a thick rootstock; milky latex present; plants monoecious.
STEMS: Erect, branching; armed with stiff, stinging, glandular-based hairs, 0.5–1.0 cm long.
LEAVES: Simple, alternate, pubescence similar to stems; blades deeply 3–5-lobed; venation palmate; petioles present.
INFLORESCENCE: Terminal; cymose; pistillate flowers few.
    CALYX: Sepals 5, lobed, fragrant, white, about 3 cm long, densely hispid.
    COROLLA: Petals absent.
    STAMINATE FLOWERS: Stamens 10, connate, included within the throat of the calyx; with soft pubescence at the base of the filaments.
    PISTILLATE FLOWERS: Occupying the central position in the inflorescence; ovary superior, hispid; styles 3, each dichotomously branched.
FRUIT: A 3-seeded capsule.
COMMENTS: Common on deep sandy soils thoughout the Rio Grande Plains and Coastal Prairies. The seeds are eaten by Rio Grande turkeys and mourning doves.

EUPHORBIACEAE

Croton L.
1a. Plants dioecious. **Croton texensis**
1b. Plants monoecious. **2**

2a. Leaf margins toothed. **Croton glandulosus**
2a. Leaf margins entire. **3**

3b. Mature capsules bearing 3 seeds; stamens 7–11. **Croton capitatus**
3a. Mature capsules bearing 1 seed; stamens 4–5. **Croton monanthogynous**

WOOLLY CROTON, HOGWORT
Croton capitatus Michx. var. **lindheimeri** (Engelm. & Gray) Muell. Arg.

ANNUAL: From a taproot; plants monoecious.
STEMS: Erect, branching at the nodes; grayish-green; stellate pubescent.
LEAVES: Simple, alternate, stellate, silvery pubescent; blades ovate-lanceolate; base cordate to subcordate; margins entire; apex acute or obtuse; petioles and stipules present.
INFLORESCENCE: A raceme; males located above the females.
STAMINATE FLOWERS
    CALYX: Sepals 5, inconspicuous; stellate-pubescent.
    COROLLA: Petals 5, inconspicuous; stellate-pubescent.
    STAMENS: 7–11; anthers white- or cream-colored.
PISTILLATE FLOWERS
    CALYX: With 6–9 lobes; stellate-pubescent.
    COROLLA: Petals absent.
    PISTIL: Ovary superior, densely pubescent; styles 3, each lobed.
FRUIT: A nearly globose, 3-seeded capsule.
COMMENTS: Common in sandy prairies, openings, and waste places thoughout the Rio Grande Plains and Coastal Prairies. The seeds are an important food of bobwhite quail, mourning doves, and Rio Grande turkeys.

EUPHORBIACEAE

NORTHERN CROTON
Croton glandulosus L. var. *septentrionalis* Muell. Arg.

ANNUAL: From a taproot; plants monoecious.
STEMS: Stellate-pubescent.
LEAVES: Simple, alternate, stellate-pubescent; blades lanceolate; margins toothed; apex obtuse and toothed.
INFLORESCENCE: In few-flowered clusters; axis pubescent.
STAMINATE FLOWERS
    CALYX: Sepals 5, minute.
    COROLLA: Absent.
    STAMENS: 10, free.
PISTILLATE FLOWERS
    CALYX: Sepals 5, linear, stellate-pubescent.
    COROLLA: Absent.
    PISTIL: Ovary superior, stellate-pubescent, 3-lobed; styles 3, each with 2 branches.
FRUIT: A capsule with 3 brown seeds; seeds about 3.5 mm long.
COMMENTS: Frequent on sandy loam, clay, and caliche soils in prairies and openings in the Rio Grande Plains and Coastal Prairies. The seeds are eaten by bobwhite quail and mourning doves.

ONE-SEEDED CROTON, PRAIRIE TEA
Croton monanthogynous Michx.

ANNUAL: From a taproot; plants monoecious.
STEMS: Erect, branching above, to 45 cm tall, with scattered, appressed, stellate pubescence.
LEAVES: Simple, alternate, grayish-green, pubescence similar to stems; blades oblong-ovate, elliptic; margins entire; apex acute; petioles present.
INFLORESCENCE: Flowers in small axillary clusters.
STAMINATE FLOWERS
    CALYX: Sepals 4–5, densely pubescent.
    COROLLA: Petals 4–5.
    STAMENS: 4 or occasionally 5.
PISTILLATE FLOWERS
    CALYX: Sepals 5, pubescent, about 1/2 length of the ovary.
    COROLLA: Absent.
    PISTIL: Ovary superior, pubescent; styles 2, each with 2 branches.
FRUIT: A single-seeded capsule plus 1 aborted ovule in immature capsules; seeds brown or amber.
COMMENTS: Common on heavier loams, clay, or caliche soils in prairies, openings, and waste places in the Rio Grande Plains and western portion of the Coastal Prairies. The seeds are frequently eaten by bobwhite quail, scaled quail, and mourning doves.

EUPHORBIACEAE

TEXAS CROTON, TINAJERA
Croton texensis (Klotzch) Muell. Arg.

ANNUAL: From a taproot; plants dioecious.
STEMS: Erect, branching above; stellate-pubescent.
LEAVES: Simple, alternate, stellate-pubescent; blades lanceolate to linear-lanceolate, to 12 cm long and 3 cm wide; margins entire; base rounded to subcordate; margins entire; apex obtuse or rounded; petioles elongated.
INFLORESCENCE: A few-flowered raceme.
STAMINATE FLOWERS
    CALYX: Sepals 5.
    COROLLA: Absent.
    STAMENS: 8–12.
PISTILLATE FLOWERS
    CALYX: Sepals 5, broadly triangular, stellate-pubescent.
    COROLLA: Absent.
    PISTIL: Ovary superior, stellate-pubescent; styles 3, each branched several times.
FRUIT: A 3-lobed capsule, each carpel with a single, large seed.
COMMENTS: Abundant on deep sand and sandy loam soils in the Rio Grande Plains and Coastal Prairies. The seeds are frequently eaten by bobwhite quail and mourning doves.

EUPHORBIACEAE

Euphorbia L.
1a. Stems often rooting at the nodes; seeds with a white substance on the seed coat, lacking transverse ridges. *Euphorbia albomarginata*
1b. Stems not rooting at the nodes; seeds with transverse ridges. *Euphorbia glyptosperma*

WHITE MARGIN EUPHORBIA
Euphorbia albomarginata T.&G.
[Syn. *Chamaesyce albomarginata* (T.&G.) Small]

PERENNIAL: Plants monoecious; all parts with a milky latex.
STEMS: Mat-forming, rooting at some nodes; glabrous and slightly ribbed.
LEAVES: Simple, opposite, glabrous; blades broadly rounded; margins entire; stipules triangular, whitish; petioles 8–10 mm long.
INFLORESCENCE: A petaloid cyathium; gland appendages white.
STAMINATE FLOWERS: 15–25 per cyathium.
PISTILLATE FLOWERS: Ovary superior; styles 3.
FRUIT: A glabrous, 3-lobed capsule; seeds about 1.5 mm long with a white substance on the seed coat.
COMMENTS: Common on clay soils in prairies, openings, and fields in the southern portion of the Rio Grande Plains and Coastal Prairies. The leaves are frequently eaten by white-tailed deer.

RIDGESEED EUPHORBIA
Euphorbia glyptosperma Engelm.
[Syn. *Chamaesyce glyptosperma* (Engelm.) Small]

ANNUAL: From a shallow taproot; all parts with a milky latex.
STEMS: Prostrate, mat-forming but not rooting at the nodes.
LEAVES: Simple, opposite, glabrous; blades rounded at the base; margins of some with irregularly spaced teeth; apex rounded; petioles present; stipules scalelike, white.
INFLORESCENCE: A cyathium.
STAMINATE FLOWERS: 4 glandular, staminate groups; glands each bearing a minute petaloid fringe; flowers with 1 stamen.
PISTILLATE FLOWERS: Ovary superior; styles 3, each minutely branched.
FRUIT: A glabrous, 3-lobed capsule; seeds with transverse ridges.
COMMENTS: Frequent on sandy soils in openings, prairies, and waste places in the Rio Grande Plains and Coastal Prairies. White-tailed deer frequently eat the leaves of this small forb, and the seeds are utilized by bobwhite quail, scaled quail, and Rio Grande turkeys.

EUPHORBIACEAE

BERLANDIER NETTLE SPURGE, JICAMILLA
Jatropha cathartica Teran & Berl.

PERENNIAL: From a large tuberlike root, some softball-sized or larger; plants monoecious.
STEMS: Developing after summer rains; erect and branching; gray-green, glabrous.
LEAVES: Simple, alternate, glabrous; blades deeply palmately dissected with 5–7 lobes; each lobe deeply cleft; petioles elongated; stipules present.
INFLORESCENCE: Cymose.
STAMINATE FLOWERS
    CALYX: Sepals 5, united, glabrous; apex mucronate.
    COROLLAS: Petals 5, free, crimson.
    STAMENS: 7–8, free, of 2 lengths; anthers yellow and red, with glands near the base.
PISTILLATE FLOWERS
    CALYX and COROLLA: Similar to staminate flowers.
    PISTIL: Ovary superior; styles 3, bifid at the apex.
FRUIT: A glabrous capsule with usually 2 seeds per segment; seeds rough-surfaced.
COMMENTS: Infrequent on clay soils in the Rio Grande Plains. The roots are eaten by javelina, and the seeds by mourning doves, bobwhite quail, and scaled quail.

KNOTWEED LEAF FLOWER
Phyllanthus polygonoides Spreng.

PERENNIAL: From a woody rootstock; plants usually less than 10 cm tall; plants usually monoecious.
STEMS: Woody below, herbaceous above, grayish-green, glabrous.
LEAVES: Simple, alternate, glabrous; blades oblong or lanceolate; margins entire; apex extended into a short mucro; petioles present; stipules membranous, oblanceolate, white.
INFLORESCENCE: Usually 1–2 flowers arising from leaf axils.
STAMINATE FLOWERS: Sepals 6; petals absent; stamens 3, about 1 mm long.
PISTILLATE FLOWERS: Sepals 6, united, glabrous with white margins; ovary superior, styles 3, bifid at apex.
FRUIT: A capsule with 2, rough-surfaced seeds per segment.
COMMENTS: Common on sandy loam or rocky caliche soils in the Rio Grande Plains. The seeds are eaten by bobwhite quail, mourning doves,and white-winged doves, and the leaves are consumed by white-tailed deer.

EUPHORBIACEAE

QUEEN'S DELIGHT, QUEEN'S ROOT
Stillingia sylvatica L.

PERENNIAL: From a woody rootstock; all parts with a milky latex; plants monoecious.
STEMS: Erect, glabrous.
LEAVES: Simple, alternate, glabrous; blades oblanceolate; margins minutely toothed; petioles present.
INFLORESCENCE: A spike with the male flowers above the females.
STAMINATE FLOWERS: With green, cuplike glands subtending a cluster of minute, green flowers.
    CALYX: 2-lobed; united below.
    COROLLA: Absent.
    STAMENS: 2; anthers yellow
PISTILLATE FLOWERS: Individual flowers subtended by 2 green, cuplike glands.
    CALYX: Scalelike; minute.
    COROLLA: Absent.
    PISTIL: Ovary superior; style 1, with 3 branches near the apex.
FRUIT: A 3-lobed capsule, each containing a large seed; seeds with a prominent caruncle.
COMMENTS: Frequent in deep sands in prairies and openings in the Rio Grande Plains and Coastal Prairies. The seeds are an important food of bobwhite quail and mourning doves.

BRUSH NOSEBURN
Tragia glanduligera Pax. & K. Hoffm.

PERENNIAL: Plants monoecious.
STEMS: Usually vinelike, twining, climbing; with irritating hairs either short and recurved or longer and sharp pointed.
LEAVES: Simple, alternate, pubescent,with irritating hairs; blades ovate or elliptic; toothed on the margins; petioles pubescent.
INFLORESCENCE: A raceme; male flowers above and females below; axis pubescent with gland-tipped hairs.
STAMINATE FLOWERS: Calyx lobes 3; stamens 3.
PISTILLATE FLOWERS: Calyx lobes 6; corolla absent.
PISTIL: Ovary superior; style 1, with 3 branches.
FRUIT: A hispid, 3-lobed capsule; seeds 1 per carpel, round, brown, with golden droplets on the surface (possibly eliasomes).
COMMENTS: Frequent in dry sandy soil in the Rio Grande Plains and Coastal Prairies. The seeds are eaten by bobwhite quail, scaled quail, and mourning doves, whereas the leaves are occasionally eaten by white-tailed deer.

FABACEAE (Leguminosae)

1a. Leaves simple. **2**
1b. Leaves compound. **3**

2a. Leaves linear or narrowly lanceolate, with a prominent vein parallel to each margin; corolla purple. *Galactia*
2b. Leaves broadly rounded; venation palmate; corolla yellow. *Rhynchosia*

3a. Leaves palmately or bipinnately compound. **4**
3b. Leaves pinnately compound or trifoliolate. **10**

4a. Leaves palmately compound; corolla papilionaceous. **5**
4b. Leaves bipinnately compound; corolla regular or slightly irregular, but not papilionaceous. **6**

5a. Leaflets 4; corolla yellow; flowers subtended by 2 ovate bracts; plants trailing or reclining; perennials. *Zornia*
5b. Leaflets 5 or more; corolla blue and white; flowers not subtended by bracts; plants erect annuals. *Lupinus*

6a. Inflorescence a raceme; leaves and perianth glandular-pubescent. *Hoffmanseggia*
6b. Inflorescence a head; leaves and perianth not glandular-pubescent. **7**

7a. Heads few-flowered, not globose, cream-colored or white; legume glabrous. *Desmanthus*
7b. Heads with numerous flowers, globose, yellow, pink or rose; legume pubescent or spiny. **8**

8a. Heads yellow. *Neptunia*
8b. Heads pink, cream-colored, or white. **9**

9a. Legumes linear, with recurved prickles; peduncle with recurved prickles. *Schrankia*
9b. Legume oblong, pubescent; peduncle pubescent but not spiny. *Mimosa*

FABACEAE (Leguminosae)

10a. Tendrils present. *Vicia*
10b. Tendrils absent. **11**

11a. Leaves even-pinnately compound. **12**
11b. Leaves odd-pinnately compound. **13**

12a. Corolla blue, papilionaceous; stamens diadelphous. *Astragalus*
12b. Corolla yellow, slightly zygomorphic; stamens distinct. *Chamaecrista*

13a. Leaves trifoliolate. **14**
13b. Leaves with more than 3 leaflets. **17**

14a. Corollas yellow. **15**
14b. Corollas not yellow. **16**

15a. Fruits burlike, surrounded with numerous hooked barbs. *Medicago*
15b. Fruits not burlike. *Rhynchosia*

16a. Corollas white or light pink; inflorescence a head; plants rooting at the nodes, mat-forming, glabrous. *Trifolium*
16b. Corollas purple or rose; inflorescence a few-flowered raceme; plants trailing, but not rooting at the nodes, glabrous. *Galactia*

17a. Inflorescence a spike; plants usually with a strong odor; leaves glandular-dotted. *Dalea*
17b. Inflorescence a raceme or spicate raceme. **18**

18a. Plants annuals; corolla blue near the apex. *Astragalus*
18b. Plants perennials; corolla salmon-red or rose-purple and white below. **19**

19a. Leaves with up to 9 leaflets; corolla salmon-red, glabrous. **Indigofera**
19b. Leaves with up to 15 leaflets; corolla rose-purple and white below; pubescent on the outer surface. *Tephrosia*

FABACEAE (Leguminosae)

BRAZOS MILK-VETCH
Astragalus brazoensis Buckl.

ANNUAL: From a taproot.
STEMS: Low-growing; appressed-pubescent.
LEAVES: Odd- or occasionally even-pinnately compound; alternate, pubescent with appressed hairs; leaflets elliptic or ovate-elliptic; margins entire; petioles and rachis with appressed hairs; stipules present.
INFLORESCENCE: A raceme from the upper leaf axils; peduncle pubescent.
    CALYX: Sepals 5, united, with appressed pubescence.
    COROLLA: Petals 5, papilionaceous, blue near the apex.
    STAMENS: 10, diadelphous.
    PISTIL: Ovary superior; style 1, unbranched.
FRUIT: A legume, broader than long.
COMMENTS: Locally abundant on sand or clay soils in prairies, openings, and waste places in the Rio Grande Plains and extreme western Coastal Prairies. The leaves are eaten by white-tailed deer, and the seeds are occasionally eaten by mourning doves.

FABACEAE (Leguminosae)

Chamaecrista (Breyne) Moench
1a. Peduncle 2.5–3.3 cm long. *Chamaecrista flexuosa*
1b. Peduncle 2 cm long or less. *Chamaecrista fasciculata*

PARTRIDGE PEA, PRAIRIE SENNA
Chamaecrista fasciculata (Michx.) Greene
[Syn. *Cassia fasciculata* Michx. var. *ferrisiae* (Britt. & Rose) B.L. Turner]

ANNUAL: From a taproot; long-lived.
STEMS: Branching extensively; appressed pubescent.
LEAVES: Even-pinnately compound, alternate, pubescent; leaflets 12–14 pairs per leaf; apex mucronate; petioles swollen near the base, each with a stalked gland near the base; stipules pubescent, herbaceous; apex acute or acuminate.
INFLORESCENCE: Flowers axillary, 1-several per node; peduncles 2 cm or less long.
    CALYX: Sepals 5, free, minutely pubescent on the back and slightly keeled, yellowish-green.
    COROLLA: Petals 5, yellow, slightly zygomorphic, clawed and reddish-tinged near the base.
    STAMENS: 10, free; anthers purple to black, with apical pores.
    PISTIL: Ovary superior, pubescent; style 1, unbranched.
FRUIT: A flattened legume.
COMMENTS: Common on sandy soils in prairies, openings, and waste places in the Coastal Prairies and Rio Grande Plains. The seeds are eaten by bobwhite quail, and the leaves are occasionally consumed by white-tailed deer.

FABACEAE (Leguminosae)

TEXAS SENNA
Chamaecrista flexuosa (L.) Greene var. *texana* (Buckley) Irwin & Barneby
[Syn. *Cassia texana* Buckl.]

PERENNIAL: From a woody rootstock.
STEMS: Trailing; densely pubescent; hairs bent and/or curly.
LEAVES: Even-pinnately compound, alternate, leaflets linear, 10–15 pairs per leaf, to 1.5 mm wide; stipules with long, straight hairs.
INFLORESCENCE: Flowers solitary in the leaf axils; peduncles 2.5–3.3 cm long, pubescent.
    CALYX: Sepals 5, free, yellow-green.
    COROLLA: Petals 5, slightly zygomorphic, golden-yellow, clawed and reddish-tinged near the base.
    STAMENS: 10, free; anthers yellow, elongated.
    PISTIL: Ovary superior, pubescent; style 1, unbranched.
FRUIT: A legume.
COMMENTS: Frequent on well-drained sand or caliche soils in pasture openings and waste places on the Rio Grande Plains. The seeds are eaten by bobwhite quail, and the foliage is occasionally eaten by cattle and white-tailed deer.

FABACEAE (Leguminosae)

Dalea L.
1a. Corollas yellow. **2**
1b. Corollas purple or rose-purple. **3**

2a. Spikes 2.0–2.5 cm wide; plants erect. ***Dalea aurea***
2b. Spikes about 1 cm wide; plants sprawling. ***Dalea nana***

3a. Plants annuals; leaflets 13–17 per leaf. ***Dalea emarginata***
3b. Plants perennials; leaflets 5–7 per leaf. ***Dalea pogonathera***

GOLDEN DALEA
Dalea aurea Nutt. ex Pursh

PERENNIAL: From a woody taproot; plants with a strong odor.
STEMS: Erect, grayish-green with appressed pubescence.
LEAVES: Odd-pinnately compound with 3–5 leaflets, alternate, pubescent; leaflets 5-7 mm wide; stipules present.
INFLORESCENCE: A terminal, conelike spike, 2.0–2.5 cm wide.
    CALYX: Sepals 5, subulate, silky pubescent; apex acuminate.
    COROLLA: Petals 5, yellow, papilionaceous.
    STAMENS: 10, connate; anthers yellow.
    PISTIL: Ovary superior, pubescent; style 1, unbranched.
FRUIT: An inconspicuous, 1-seeded, podlike legume.
COMMENTS: Occasional on deep sands in prairies and openings in the Rio Grande Plains and western Coastal Prairies. The leaves and flowers are eaten by white-tailed deer.

FABACEAE (Leguminosae)

PRAIRIE CLOVER, DALEA
Dalea emarginata (T.&G.) Shinners
[Syn. *Petalostemum emarginatum* T.&G.]

ANNUAL: From a taproot; plants with a strong odor.
STEMS: Nearly erect, glabrous.
LEAVES: Odd-pinnately compound with 13–17 leaflets, alternate, glabrous; leaflets slightly notched at the apex; stipules present.
INFLORESCENCE: An elongated spike; axis with scattered glands, 7–20 cm long; bracts at the base of the inflorescence subulate, glandular; flowers subtended by bracts.
    CALYX: Sepals 5, united, densely pubescent, apex subulate.
    COROLLA: Petals 5, rose-purple, papilionaceous.
    STAMENS: 5.
    PISTIL: Ovary superior, pubescent; style 1, unbranched.
FRUIT: A small, pubescent legume.
COMMENTS: Frequent on sandy soils in the Rio Grande Plains and Coastal Prairies. The leaves are eaten by white-tailed deer and cattle.

DALEA, PRAIRIE CLOVER
Dalea nana Torr. ex Gray

PERENNIAL: From a woody base; plants with a strong odor.
STEMS: Sprawling, densely pubescent.
LEAVES: Odd-pinnately compound with 5 leaflets, pubescent; leaflets ovate-elliptic; stipules present.
INFLORESCENCE: A compact spike, to 1 cm wide; bracts subtending flowers purple above.
    CALYX: Flowers subtended by an ovate bract, silky-villous, apex mucronate; sepals 5, united, apex acuminate, villous.
    COROLLA: Petals 5, yellow- or copper-colored, papilionaceous.
    STAMENS: 10, connate.
    PISTIL: Ovary superior, pubescent; style 1, unbranched.
FRUIT: A pubescent, podlike legume.
COMMENTS: Frequent on sandy loam and caliche soils in prairies and openings in the Rio Grande Plains and Coastal Prairies. The leaves are eaten by white-tailed deer.

FABACEAE (Leguminosae)

BEARDED DALEA, HIERBA DEL CORAZON
Dalea pogonathera Gray

PERENNIAL: From a woody caudex.
STEMS: Glabrous with scattered, glandular spots.
LEAVES: Odd-pinnately compound with 5–7 leaflets, alternate, densely glandular-punctate; leaflets linear, about 4 mm long; apex acuminate.
INFLORESCENCE: A spike; each flower subtended by a white, glandular, acuminate bract.
    CALYX: Sepals 5, united, silky villous, apex acuminate, brown.
    COROLLA: Petals 5, purple, clawed at the base, papilionaceous; lateral petals often with a green base.
    STAMENS: 10, connate.
    PISTIL: Ovary superior; style 1, unbranched.
FRUIT: A densely villous legume.
COMMENTS: Frequent on sandy loam and caliche soils in the Rio Grande Plains. The leaves and flowers are eaten by white-tailed deer.

BUNDLEFLOWER
Desmanthus virgatus (L.) Willd. var. *depressus* (Willd.) B.L. Turner
[Syn. *Desmanthus depressus* Willd.]

PERENNIAL
STEMS: Sprawling or scandent; glabrous, angled.
LEAVES: Bipinnately compound, alternate; leaflets glabrous but margins with scattered cilia, 2.5–3.0 mm long; stipules present; glands present at base of pinna.
INFLORESCENCE: Few-flowered heads from leaf axils; bracts subtending flowers membranous, brownish with an acuminate apex.
    CALYX: Sepals 5, united; apex mucronate.
    COROLLA: Petals 5, green with white margins, actinomorphic.
    STAMENS: 10, anthers yellow.
    PISTIL: Ovary superior; style 1, unbranched.
FRUIT: A linear legume, 3.5–6.0 cm long.
COMMENTS: Common on various soils in prairies, openings, and waste places in the Rio Grande Plains and Coastal Prairies. The foliage is eaten by white-tailed deer and cattle, and the seeds are consumed by bobwhite quail, scaled quail, and mourning doves.

FABACEAE (Leguminosae)

Galactia P. Br.
1a. Leaves simple. *Galactia marginalis*
1b. Leaves trifoliolate. *Galactia canescens*

HOARY MILKPEA
Galactia canescens Benth.

PERENNIAL
STEMS: Prostrate, trailing, but not twining; appressed-pubescent.
LEAVES: Trifoliolate, alternate, appressed-pubescent; leaflets round to elliptic, the terminal leaflet 1–3 cm long, grayish; margins entire; petioles pubescent; stipules present.
INFLORESCENCE: Few-flowered racemes from leaf axils.
  CALYX: Sepals 4, united, unequal.
  COROLLA: Petals 5, rose to purple, papilionaceous.
  STAMENS: 10, diadelphous.
  PISTIL: Ovary superior, pubescent; style 1, unbranched.
FRUIT: A somewhat turgid, linear, pubescent legume; seeds brown, mottled, 5–7 mm long.
COMMENTS: Frequent in sandy prairies and openings in the Rio Grande Plains and Coastal Prairies. The leaves are an important food of white-tailed deer and are also eaten by cattle. The seeds are eaten by bobwhite quail, mourning doves, and Rio Grande turkeys.

MILKPEA
Galactia marginalis Benth.

PERENNIAL
STEMS: Erect to trailing.
LEAVES: Simple, alternate, glabrous above, appressed-pubescent below; blades linear or narrowly lanceolate with a prominent vein parallel to each margin; margins entire; apex obtuse, some with a mucronate tip; stipules linear.
INFLORESCENCE: Single- or several-flowered from the leaf axils; pedicels pubescent; linear bracts present below the calyx.
  CALYX: Sepals 5, united, appressed pubescent.
  COROLLA: Petals 5, purple, papilionaceous.
  STAMENS: 10, diadelphous.
  PISTIL: Ovary superior, pubescent; style 1, unbranched.
FRUIT: A legume.
COMMENTS: Infrequent in sandy prairies and caliche areas in the eastern portion of the Rio Grande Plains and Coastal Prairies. The leaves are eaten by white-tailed deer.

FABACEAE (Leguminosae)

SICKLEPOD RUSHPEA
Hoffmanseggia glauca (Ort.) Eifert
[Syn. *Hoffmanseggia densiflora* Gray]

PERENNIAL: From rhizomes.
STEMS: Minutely pubescent, ribbed.
LEAVES: Bipinnately compound, alternate, glabrous; leaflets elliptic; margins entire, minutely pubescent; petiole and rachis remotely glandular.
INFLORESCENCE: A raceme with a glandular and pubescent axis; pedicels stalked, glandular and pubescent.
    CALYX: Sepals 5 with stalked glands.
    COROLLA: Petals 5, clawed at the base, yellow, slightly zygomorphic, with a dense covering of stalked glands.
    STAMENS: 10, free; filaments pubescent with stalked glands.
    PISTIL: Ovary superior; style 1, unbranched.
FRUIT: A legume.
COMMENTS: Occasional on sand or clay loam soils in the Rio Grande Plains and Coastal Prairies. The leaves are eaten by white-tailed deer.

WESTERN INDIGO, SCARLET PEA
Indigofera miniata Ort. var. *leptosepala* (Nutt.) B.L. Turner

PERENNIAL: From a woody rootstock.
STEMS: Trailing but not rooting at the nodes; appressed-pubescent.
LEAVES: Odd-pinnately compound with up to 9 leaflets, alternate, pubescent; leaflets somewhat V-shaped; stipules present.
INFLORESCENCE: A spicate raceme.
    CALYX: Sepals 5, united, pubescent.
    COROLLA: Petals 5, salmon-red, papilionaceous, longer than 8 mm; readily dehiscent.
    STAMENS: 10, diadelphous, connate.
    PISTIL: Ovary superior; style 1, unbranched.
FRUIT: A linear, pubescent legume with appressed hairs.
COMMENTS: Abundant on sandy loam and deep sandy soils in the Rio Grande Plains and Coastal Prairies. The leaves and stems are eaten by white-tailed deer, Rio Grande turkeys, and cattle, and the seeds are occasionally eaten by bobwhite quail and mourning doves.

FABACEAE (Leguminosae)

TEXAS BLUEBONNET
Lupinus subcarnosus Hook.

ANNUAL: From a taproot.
STEMS: Erect, densely silky-pubescent.
LEAVES: Palmately compound with usually 5 leaflets, alternate; densely pubescent; leaflets oblanceolate; margins entire; apex obtuse; petioles and stipules present.
INFLORESCENCE: A terminal raceme about 3 cm wide; pedicels pubescent.
    CALYX: Sepals 5, united, pubescent.
    COROLLA: Petals 5, blue and white; papilionaceous.
    STAMENS: 10, connate; anthers yellow.
    PISTIL: Ovary superior, pubescent; style 1, unbranched.
FRUIT: A pubescent legume.
COMMENTS: Frequent on sandy loam, sandy, and caliche soils throughout the Rio Grande Plains and Coastal Prairies. The seeds are eaten by bobwhite quail.

BUR-CLOVER, MEDIC
Medicago polymorpha L. var. *vulgaris* (Benth.) Shinners
[Syn. *Medicago hispida* Gaertn.]

ANNUAL: From a taproot; often bearing nodules.
STEMS: Erect or usually sprawling, slightly angled, glabrous.
LEAVES: Trifolioliate, alternate, glabrous leaflets, cordate to obovate, margins toothed; pedicels grooved; stipules with linear lobes.
INFLORESCENCE: 3–5 flowers arising from the upper leaf axils.
    CALYX: Sepals 5, united below, about 1/2 –2/3 length of corolla, slightly pubescent, lobes linear.
    COROLLA: Petals 5, yellow, papilionaceous.
    STAMENS: 10, diadelphous.
    PISTIL: Ovary superior; style 1, unbranched.
FRUIT: A rounded legume with numerous hooked barbs, fruit appearing burlike, style 1, unbranched.
COMMENTS: Common on disturbed sites on various soils in the Coastal Prairies and Rio Grande Plains. This plant is occasionally eaten by white-tailed deer and cattle.

FABACEAE (Leguminosae)

HERBACEOUS MIMOSA, POWDERPUFF
Mimosa strigillosa T.&G.

PERENNIAL
STEMS: Spreading, prostrate; covered with bristles.
LEAVES: Bipinnately compound, alternate, glabrous; leaflets linear, slightly curved, margins with scattered bristles; petioles and rachis prickly; stipules broadly triangular; pinna "sensitive" to the touch.
INFLORESCENCE: A globose head; peduncle elongated, with bristles.
    CALYX: Sepals 5, united; lobes triangular with a few minute hairs.
    COROLLA: Absent.
    STAMENS: 10, free; filaments pink, rose or reddish-purple; anthers cream-colored.
    PISTIL: Ovary superior; style 1, unbranched.
FRUIT: An oblong, jointed, pubescent legume.
COMMENTS: Frequent on various soils, mostly in swales, depressions and along stream or lake banks in the eastern portion of the Rio Grande Plains and Coastal Prairies. The leaves are eaten by white-tailed deer and cattle.

TROPICAL NEPTUNIA, YELLOW PUFF
Neptunia pubescens Benth.

PERENNIAL
STEMS: Sprawling or reclining; brownish-pubescent; slightly ribbed.
LEAVES: Bipinnately compound with 3–4 pairs of pinnae, alternate, densely pubescent with glandular-tipped hairs; leaflets linear, to 4 mm long; apex obtuse; petiole and pinnae with a pulvinus at their base; "sensitive" to the touch; rachis slightly-grooved; stipules veiny with an acute apex.
INFLORESCENCE: A globular head arising from the upper leaf axils; peduncle shorter than leaves, with glandular hairs.
    CALYX: Sepals 5, united.
    COROLLA: Petals 5, lemon-yellow, united below.
    STAMENS: 10; uppermost flowers can lack stamens.
    PISTIL: Ovary superior, pubescent; style 1, unbranched.
FRUIT: A flattened, pubescent legume 3.0–3.5 cm long.
COMMENTS: Frequent in swales, ditches, waste places, and along freshwater shores in the Coastal Prairies and eastern portion of the Rio Grande Plains. The seeds are eaten by bobwhite quail and mourning doves, whereas the leaves are consumed by white-tailed deer and cattle.

FABACEAE (Leguminosae)

Rhynchosia Lour.
1a. Leaves simple. ***Rhynchosia americana***
1b. Leaves trifoliolate. **2**

2a. Leaflets ovate to rounded. ***Rhynchosia minima***
2b. Leaflets lanceolate. ***Rhynchosia senna***

AMERICAN SNOUTBEAN
Rhynchosia americana (Mill.) C. Metz

PERENNIAL: From a woody rootstock.
STEMS: Trailing, but not rooting at the nodes; pubescent; variously-angled.
LEAVES: Simple, alternate, pubescent; blades broadly rounded, venation palmate, cordate at the base; apex rounded; petioles present; stipules pubescent.
INFLORESCENCE: A few flowers arising from the leaf axils.
    CALYX: Sepals 5, united, 2 V-shaped, the 3rd lobed to the basal tube, pubescent.
    COROLLA: Petals 5, yellow, papilionaceous; 3-clawed near the base.
    STAMENS: 10, diadelphous, connate.
    PISTIL: Ovary superior; style 1, unbranched.
FRUIT: A small pubescent legume.
COMMENTS: Frequent in sandy prairies and openings in the Rio Grande Plains and western portion of the Coastal Prairies. The seeds are an important food of bobwhite quail, whereas the leaves are frequently consumed by white-tailed deer and cattle.

FABACEAE (Leguminosae)

LEAST SNOUTBEAN
Rhynchosia minima (L.) DC.

PERENNIAL
STEMS: Trailing or twining, branching at the nodes; variously angled, pubescent.
LEAVES: Trifoliolate, alternate, glabrous above, pubescent on the veins below; leaflets ovate to rounded; apex rounded; stipules present.
INFLORESCENCE: A raceme from the upper leaf axils, with 5 or more flowers.
    CALYX: Sepals 5, united, unequal, pubescent.
    COROLLA: Petals 5, yellow with purple veins and pubescent on the back, papilionaceous.
    STAMENS: 10, diadelphous; anthers yellow.
    PISTIL: Ovary superior; style 1, unbranched.
FRUIT: A curved legume.
COMMENTS: Frequent on various soils in pastures, woods, along ravines, and in stream bottoms in the Coastal Prairies and Rio Grande Plains. The leaves are eaten by white-tailed deer and cattle, whereas the seeds are eaten by several species of birds.

TEXAS SNOUTBEAN
Rhynchosia senna Gillies ex Hook. var. *texana* (T.&G.) M.C. Johnston
[Syn. *Rhynchosia texana* T.&G.]

PERENNIAL
STEMS: Twining or trailing vine; pubescent.
LEAVES: Trifoliolate, alternate, slightly pubescent; leaflets lanceolate; margins entire; apex obtuse; petioles pubescent.
INFLORESCENCE: 1–3 flowers arising from the leaf axils; pedicels densely pubescent.
    CALYX: Sepals 5, united, pubescent.
    COROLLA: Petals 5, yellow, papilionaceous.
    STAMENS: 10, diadelphous.
    PISTIL: Ovary superior.
FRUIT: A legume.
COMMENTS: Frequent on sandy loam, clay, or caliche soils in prairies, pastures, and woods in the Coastal Prairies. The leaves are eaten by white-tailed deer.

FABACEAE (Leguminosae)

KARNES SENSITIVE BRIAR
Schrankia latidens (Small) K. Schum.

PERENNIAL: From a woody rootstock.
STEMS: Trailing, winged, with short, recurved prickles.
LEAVES: Bipinnately compound, alternate, glabrous; leaflets linear, glaucous; apex with a short mucro; petioles with recurved prickles; stipules with soft spines; pinna "sensitive" to the touch.
INFLORESCENCE: A globose head arising from the leaf axils; peduncle with recurved prickles.
 CALYX: Sepals 5, united about ½ their length, glabrous.
 COROLLA: Absent.
 STAMENS: Usually 10, free; filaments bright pink; anthers yellow.
 PISTIL: Ovary superior; style 1, pink, unbranched.
FRUIT: A linear legume covered with recurved prickles.
COMMENTS: Frequent on sandy soils in prairies and openings in the Rio Grande Plains and Coastal Prairies. The seeds are eaten by bobwhite quail and mourning doves, and the leaves are consumed by white-tailed deer.

LINDHEIMER TEPHROSIA, ROUNDLEAF TEPHROSIA
Tephrosia lindheimeri Gray

PERENNIAL: From a woody rootstock.
STEMS: Trailing, grayish-green, densely pubescent.
LEAVES: Odd-pinnately compound with up to 15 leaflets, alternate; leaflets ovate or elliptic; margins highlighted with white hairs; apex obtuse, with a short awn; stipules present.
INFLORESCENCE: A few-flowered raceme; peduncle densely pubescent.
 CALYX: Sepals 5, united, greenish-yellow, densely pubescent.
 COROLLA: Petals 5, rose-purple (rarely white), the banner with a white spot below; papilionaceous, pubescent on the outer surface.
 STAMENS: 10, diadelphous, connate.
 PISTIL: Ovary superior; style 1, unbranched.
FRUIT: A curved legume.
COMMENTS: Occasional on well-drained sands in prairies and openings in the Rio Grande Plains and Coastal Prairies. The leaves and stems are eaten by white-tailed deer and cattle; the seeds are eaten by bobwhite quail.

FABACEAE (Leguminosae)

WHITE SWEET CLOVER
Trifolium repens L.

PERENNIAL
STEMS: Prostrate, mat-forming, rooting at the nodes; glabrous.
LEAVES: Trifoliolate, alternate, glabrous; leaflets ovate with a conspicuous midvein; margins minutely serrate; petioles present.
INFLORESCENCE: A head on an elongated, glabrous peduncle; lowermost pedicels recurved.
    CALYX: Sepals 5, united, lobes linear, glabrous.
    COROLLA: Petals 5, white, turning pink with age, papilionaceous.
    STAMENS: 10, diadelphous.
    PISTIL: Ovary superior; style 1, unbranched.
FRUIT: A few-seeded loment.
COMMENTS: An introduced, cultivated species found scattered on sandy soils in the Coastal Prairies. The leaves are occasionally eaten by white-tailed deer, cattle, and Rio Grande turkeys.

Vicia L.
1a. Stems often shaggy-pubescent; inflorescence a few- to several-flowered axillary raceme. *Vicia leavenworthii*
1b. Stems usually only slightly pubescent; inflorescence of 1–2 flowers from the leaf axils; peduncle about one-half the length of the leaf. *Vicia ludoviciana*

DEER PEA VETCH
Vicia ludoviciana Nutt. var. *texana* (T.&G.) Shinners

ANNUAL
STEMS: A low-growing vine, shaggy-pubescent.
LEAVES: Pinnately compound with 5 or more pairs of leaflets, with a terminal tendril, alternate; leaflets linear; stipules present.
INFLORESCENCE: Flowers 1–2 from the leaf axils; peduncle one-half the length of the leaf, pubescent.
    CALYX: Sepals 5, united, of unequal lengths, pubescent; apex acuminate.
    COROLLA: Petals 5, blue and white with dark blue nectar guides, 5–6 mm long, papilionaceous.
    STAMENS: 10, diadelphous.
    PISTIL: Ovary superior; style 1, unbranched.
FRUIT: A narrow, flattened legume, 1.5–2.0 cm long.
COMMENTS: Frequent on various soils in the Rio Grande Plains and Coastal Prairies. The leaves and seeds are eaten by white-tailed deer, cattle, bobwhite quail, and Rio Grande turkeys.

FABACEAE (Leguminosae)

LEAVENWORTH VETCH
Vicia leavenworthii T. & G.

Similar to *V. ludoviciana.* However, the stems are only slightly pubescent. It is frequent on various soils in the Coastal Prairies. The leaves are eaten by white-tailed deer and cattle. [Not illustrated.]

BRACTED ZORNIA, VIPERINA
Zornia bracteata J.F. Gmel.

PERENNIAL: From a woody taproot.
STEMS: Trailing or reclining, pubescent.
LEAVES: Compound with 4 leaflets, alternate, glabrous above, pubescent below; leaflets lanceolate; margins appressed-pubescent; stipules winged near the base.
INFLORESCENCE: A loosely flowered raceme; each flower subtended by 2 green, ovate, pubescent, veiny bracts.
    CALYX: Sepals 5, united, pubescent.
    COROLLA: Petals 5, golden-yellow, much longer than the calyx, papilionaceous.
    STAMENS: 10, connate.
    PISTIL: Ovary superior; style 1, unbranched.
FRUIT: A spiny, pubescent loment with up to 7 segments.
COMMENTS: Frequent on sandy and gravelly soils in prairies and openings in the Rio Grande Plains and Coastal Prairies. The leaves are eaten by white-tailed deer.

FUMARIACEAE

SCRAMBLED EGGS
Corydalis micrantha (Engelm.) Gray var. *texensis* (G. Ownbey) Shinners

ANNUAL: From a taproot.
STEMS: Rather weak and falling, glabrous.
LEAVES: Compound with deeply dissected ultimate segments, alternate, bluish green; segments about 1–2 mm wide.
INFLORESCENCE: A raceme with linear bracts subtending the pedicels.
    CALYX: Sepals 2, free, minute, scarious.
    COROLLA: Petals 4, yellow, in 2 dissimilar sets, zygomorphic, spurred at the base; outer petals to 1.5 cm long.
    STAMENS: 6, connate.
    PISTIL: Ovary superior; style 1.
FRUIT: A linear capsule to 3.5 cm long; seeds rounded, burgundy-colored.
COMMENTS: Frequent on various soils in prairies, openings, fields, and waste places in the Rio Grande Plains and Coastal Prairies. The leaves are occasionally eaten by white-tailed deer and bobwhite quail; the seeds are eaten by Rio Grande turkeys.

GERANIACEAE

1a. Petals about 7 mm long; stems with a mixture of long and short hairs.
Geranium carolinianum
1b. Petals about 5 mm long; stems with short appressed hairs. *Geranium texanum*

WILD GERANIUM
Geranium carolinianum L.

ANNUAL or PERENNIAL
STEMS: Pubescent with a combination of long and short hairs.
LEAVES: Simple, but deeply, palmately lobed, alternate, opposite or occasionally 3 at a node, pubescent; blades rounded in outline, grayish-green, venation palmate; lobes toothed; petioles pubescent, to 15 cm long; stipules membranous; apex acute or acuminate.
INFLORESCENCE: Few-flowered, arising from the nodes.
    CALYX: Sepals 5, united at the base; lobes lanceolate or ovate, with a prominent midvein and 2 lateral veins, pubescent; margins translucent, ciliate.
    COROLLA: Petals 5, free, about 7 mm long, white or pinkish-white.
    STAMENS: 10.
    PISTIL: Ovary superior; style column about 1 cm long.
FRUIT: A capsule.
COMMENTS: Frequent on sandy soils in prairies, openings, and along roads in the Rio Grande Plains and Coastal Prairies. The leaves are eaten by white-tailed deer, cattle, and Rio Grande turkeys, and the seeds are consumed by mourning doves.

Geranium texanum (Trel.) Heller
[Syn. *Geranium carolinianum* L. var. *texanum* Trel.]

This species has stems with retrorse pubescence and petals about 5 mm long. The leaves are eaten by white-tailed deer. It is similar in appearance to *Geranium carolinianum.* [Not illustrated.]

HYDROPHYLLACEAE

1a. Flowers 1-few in the leaf axils; corolla pink or purple with a yellow throat. **Nama**
1b. Flowers in a scorpioid cyme; corollas blue, often with a white center. **Phacelia**

ROUGH NAMA, SANDBELL
Nama hispidum Gray

ANNUAL: From a taproot.
STEMS: Erect, hispid.
LEAVES: Simple, alternate, hirsute; blades lanceolate to oblanceolate; subsessile.
INFLORESCENCE: 1-few flowers in the leaf axils.
  CALYX: Sepals 5, free, linear, pubescent.
  COROLLA: Petals 5, united into a tube, pink or purple, yellow in the tube near the base.
  STAMENS: 5, 3 short and 2 long, epipetalous, included within the corolla.
  PISTIL: Ovary superior; styles 2.
FRUIT: A small capsule.
COMMENTS: Frequent on sandy soils in prairies and openings in the Rio Grande Plains and Coastal Prairies. The leaves are occasionally eaten by white-tailed deer and cattle.

HYDROPHYLLACEAE

Phacelia Juss.

1a. Leaves pinnately dissected; corolla 3–6 mm broad. *Phacelia congesta*

1b. Leaves only lobed, pinnatifid, or broadly-toothed; corolla 8–20 mm broad. *Phacelia patuliflora*

SPIKE PHACELIA, BLUE CURLS
Phacelia congesta Hook.

ANNUAL: From a taproot.
STEMS: Erect, branching at the nodes; densely soft-pubescent with hairs of different sizes, slightly viscid.
LEAVES: Simple, alternate, densely pubescent; pinnatifid, lobes of various dimensions; petioles present.
INFLORESCENCE: A tightly coiled, scorpioid cyme.
  CALYX: Sepals 5, united near the base; lobes lanceolate, 3–4 mm long, pubescent.
  COROLLA: Petals 5, blue; united, lobes 3–4 mm long, actinomorphic.
  STAMENS: 5, epipetalous; filaments blue; anthers white, turning brown.
  PISTIL: Ovary superior, pubescent; style 1, branches 2, pubescent below.
FRUIT: A capsule.
COMMENTS: Frequent on sandy soils in prairies and openings in the Rio Grande Plains and Coastal Prairies. The leaves are occasionally eaten by white-tailed deer, cattle, bobwhite quail, and Rio Grande turkeys.

SAND PHACELIA
Phacelia patuliflora (Engelm. & Gray) Gray var. *austrotexana* J.A Moyer

ANNUAL: From a taproot.
STEMS: Pubescent.
LEAVES: Simple, alternate, sessile, pubescent; blades oblong to ovate; margins lobed or broadly toothed.
INFLORESCENCE: A scorpioid cyme.
  CALYX: Sepals 5, united, ciliate.
  COROLLA: Petals 5, united, lobes blue or lavender, white at the base, actinomorphic.
  STAMENS: 5, epipetalous, elongated above the corolla; filaments pubescent near the base.
  PISTIL: Ovary superior, pubescent; style 1, branches 2.
FRUIT: A capsule.
COMMENTS: Frequent on sandy soils in the eastern portion of the Rio Grande Plains and Coastal Prairies. The leaves are known to be eaten by white-tailed deer.

IRIDACEAE

1a. Tepals dark purple to reddish-purple; leaves pleated; aerial stem arising from a bulb. **Alophila**
1b. Tepals blue, yellow at the base; leaves grasslike, not pleated; bulb absent. **Sisyrinchium**

PURPLE PLEATLEAF, PRAIRIE NYMPH
Alophila drummondii (Graham) Foster
[Syn. **Eustylis purpurea** (Herb.) Engelm. & Gray] [Syn. **Herbertia drummondii** Graham]

PERENNIAL: From a firm, brown, scaly bulb.
STEMS: Aerial stem erect, glabrous.
LEAVES: Simple, alternate, glabrous; blades linear-lanceolate, venation parallel, pleated with elevated veins; petioles absent; sheath present.
INFLORESCENCE: Flowers emerging from a spathe.
  TEPALS: Perianth with 3 outer and 3 inner segments; the inner depressed and forming a shallow cup; segments usually dark purple to reddish-purple, about 2.5 cm long; apex of inner tepals fringed.
  STAMENS: 3; anthers yellow.
  PISTIL: Ovary inferior; style 1, branched near the apex.
FRUIT: A many-seeded capsule; seeds orange or rust-colored.
COMMENTS: Occasional on sandy soils in the Rio Grande Plains and Coastal Prairies. The leaves are eaten by sandhill cranes.

BERMUDA BLUE-EYE GRASS
Sisyrinchium angustifolium Mill.
[Syn. **Sisyrinchium bermudiana** L.]

PERENNIAL: From fibrous roots.
STEMS: Glabrous with lateral wings wider than the stem axis.
LEAVES: Simple, tufted at the base, alternate above, glabrous; blades linear, grasslike, venation parallel; margins occasionally minutely pubescent.
INFLORESCENCE: Few-flowered, arising from a spathe-like structure near the stem apex; pedicels minutely pubescent, about 1 cm long.
  TEPALS: 6, united below, blue with a yellow throat, actinomorphic.
  STAMENS: 3, adnate to the inner tepals; anthers yellow-orange.
  PISTIL: Ovary inferior; style 1.
FRUIT: A glabrous, 3-lobed capsule.
COMMENTS: Frequent on moist sand and clay in the eastern portion of the Rio Grande Plains and Coastal Prairies. The leaves are occasionally eaten by white-tailed deer, bobwhite quail, and Rio Grande turkeys.

KRAMERIACEAE

TRAILING RATANY
Krameria lanceolata Torr.

PERENNIAL: From a woody rootstock.
STEMS: Trailing, densely pubescent.
LEAVES: Simple, alternate, sessile or subsessile, pubescent; blades lanceolate or linear; margins entire.
INFLORESCENCE: Flowers solitary; pedicel subtended by paired, pubescent, leafy bracts.
  CALYX: Sepals 5, free, reddish-purple, unequal, pubescent on the back, to 1.5 cm long.
  COROLLA: Petals 5, 3 united, red, extended above the pistil; 2 laterals free, reddish-tinged and ear-shaped (highly zygomorphic).
  STAMENS: 4, free or adnate to the base of the upper petal; anthers white.
  PISTIL: Ovary superior, densely pubescent; style 1.
FRUIT: An indehiscent, rounded, pubescent, 1-seeded pod, armed with sharp prickles 2–4 mm long.
COMMENTS: Occasional on various soils in the Rio Grande Plains and Coastal Prairies. The leaves are eaten by white-tailed deer.

LAMIACEAE (Labiatae)

1a. Stamens 2. **2**
1b. Stamens 4. **3**

2a. Corollas crimson; leaves glabrous; plants without a noticeable odor. *Salvia*
2b. Corollas greenish-white or white with purple spots; plants with a strong minty odor. *Monarda*

3a. Calyx with 1 lobe bearing a prominent ridge; corollas blue with white spots on the lower lip; plants perennials. *Scutellaria*
3b. Calyx not as above; corollas pink, purple or lavender; plants annuals. **4**

4a. Inflorescence axillary with 2–3 flowers per node; calyx lobes nearly equal. *Stachys*
4b. Inflorescence a spicate raceme or a spicate-panicle; calyx lobes unequal. *Brazoria*

Brazoria Engelm. & Gray
1a. Inflorescence an uninterrupted spicate raceme, internodes between flowers up to 15 cm. *Brazoria arenaria*
1b. Inflorescence usually a spicate-panicle; internodes inconspicuous within the inflorescence. *Brazoria truncata*

SAND BRAZORIA
Brazoria arenaria Lundell

ANNUAL: From a taproot.
STEMS: Erect, 4-angled, pubescent with both glandular and aglandular hairs.
LEAVES: Simple, opposite, mostly glabrous or with a few scattered hairs; blades oblanceolate; margins crenate.
INFLORESCENCE: An elongated, interrupted raceme with inconspicuous bracts on the axis; internodes between the flowers to 1.5 cm long.
    CALYX: Sepals 5, zygomorphic; united, the upper 3 with teeth on the lobes, the lower limb 2-lobed and toothed at the apex; pubescent with glandular and aglandular hairs; calyx enlarged in fruit.
    COROLLA: Petals 5, zygomorphic; bilabiate, with the upper lip 2-lobed and the lower 3-lobed; pink and white or purple and white; slightly pubescent on the inner throat; pubescent within the tube.
    STAMENS: 4, epipetalous; 2 long and 2 short.
    PISTIL: Ovary superior; style 1, with 2 short branches.
FRUIT: 4 pubescent, bony nutlets.
COMMENTS: Occasional on sandy soils in the Rio Grande Plains and Coastal Prairies. The leaves are eaten by white-tailed deer.

LAMIACEAE (Labiatae)

BLUNTSEPAL BRAZORIA, RATTLESNAKE FLOWER
Brazoria truncata (Benth.) Engelm. & Gray

ANNUAL: From a taproot.
STEMS: Erect, 4-angled, slightly pubescent.
LEAVES: Simple, opposite, mostly glabrous, but slightly puntate-glandular, lower leaves with elongated petioles, the uppermost leaves sessile; blades oblanceolate to nearly lanceolate; margins entire or remotely toothed.
INFLORESCENCE: An elongated, spicate-panicle; the axis pubescent.
    CALYX: Sepals 5, zygomorphic, united, the upper 3 with teeth on the lobes, the lower limb 2-lobed and toothed at the apex, pubescent with mainly strigose hairs.
    COROLLA: Petals 5, zygomorphic, bilabiate with the upper lip 2-lobed and the lower 3-lobed; lavender, the inner throat light yellow with purple spots.
    STAMENS: 4, epipetalous, 2 long and 2 short.
    PISTIL: Ovary superior; style 1, with 2 short branches.
FRUIT: 4 pubescent, bony nutlets.
COMMENTS: Locally abundant on deep sands in prairies and openings in the Coastal Prairies. The leaves are eaten by white-tailed deer.

LAMIACEAE (Labiatae)

Monarda L.

1a. Plants annuals; bracts and perianth purple-tinged. *Monarda citriodora*
1b. Plants perennials; bracts and perianth greenish-white. ***Monarda punctata***

LEMON BEEBALM, LEMON MINT, HORSEMINT
Monarda citriodora Cerv. ex Lag.

ANNUAL: From a taproot; with a strong, minty odor.
STEMS: Erect, often with short lateral branches at the nodes; pubescent, ciliate at the nodes.
LEAVES: Simple, whorled, pubescent, punctate-pitted below; blades oblong or lanceolate; margins ciliate and irregularly toothed; petioles ciliate.
INFLORESCENCE: In sessile axillary clusters; flowers subtended by bracts of various dimensions, the larger purple-tinged; margins ciliate with perpendicular hairs.
    CALYX: Sepals 5, united, tubular; oil droplets usually present on the outer surface; lobes needlelike, about 5 mm long; apex ciliate.
    COROLLA: Petals 5, zygomorphic, bilabiate, white; purple-maculate on the lower lips; ciliate with oil droplets usually present on the outer surface, about 2 cm long.
    STAMENS: 2, epipetalous; anthers purple.
    PISTIL: Ovary superior; style 1, irregularly branched near the apex.
FRUIT: 4, oblong, bony nutlets.
COMMENTS: Common on loams and clays in prairies, openings, and waste places in the Rio Grande Plains and Coastal Prairies. The leaves and bracts are eaten by white-tailed deer and cattle.

SPOTTED BEEBALM
Monarda punctata L.

PERENNIAL: With a strong, minty odor.
STEMS: Erect, 4-angled, densely pubescent.
LEAVES: Simple, opposite, pubescent and glandular punctate; blades lanceolate to broadly lanceolate; margins toothed; petioles present.
INFLORESCENCE: In axillary spikes subtended by greenish-white bracts.
    CALYX: Sepals 5, united, lobes triangular, pubescent, to 1.5 mm long.
    COROLLA: Petals 5, zygomorphic, bilabiate, greenish-white, pubescent on the outside with punctate spots, about 1.5 cm long.
    STAMENS: 2, epipetalous.
    PISTIL: Ovary superior; style 1, branches 2.
FRUIT: 4, bony nutlets.
COMMENTS: Frequent on sandy soils in prairies and openings in the Rio Grande Plains and Coastal Prairies. The leaves and bracts are eaten by white-tailed deer.

LAMIACEAE (Labiatae)

TROPICAL SAGE, MIRTO, TEXAS SAGE
Salvia coccinea Buchoz ex Etlinger

PERENNIAL: Slightly woody below.
STEMS: Erect, 4-angled, pubescent.
LEAVES: Simple, opposite, glabrous; blades broadly ovate to nearly triangular, venation palmate; margins crenate; petioles present.
INFLORESCENCE: In verticels at the nodes subtended by lanceolate, pubescent bracts; pedicels pubescent.
    CALYX: Sepals 5, united about 3/4 their length; lobes with irregular teeth; reddish-tinged.
    COROLLA: Petals 5, zygomorphic, bilabiate, scarlet, pubescent.
    STAMENS: 2, epipetalous, exserted above the corolla tube; filaments reddish, forked below.
    PISTIL: Ovary superior, reddish above; style 1, branches 2.
FRUIT: 4, bony nutlets.
COMMENTS: Frequent on loamy soils or caliche usually in partial shade in pastures and woods in the Rio Grande Plains and Coastal Prairies. The leaves are occasionally eaten by white-tailed deer.

Scutellaria L.
1a. Leaf margins entire. *Scutellaria drummondii*
1b. Leaf margins toothed. *Scutellaria muriculata*

DRUMMOND'S SKULLCAP
Scutellaria drummondii Benth.

ANNUAL: Branching at the base.
STEMS: Low-growing, branching near the base, 4-angled, densely pubescent with straight hairs.
LEAVES: Simple, opposite or nearly so, densely pubescent; blades ovate; margins entire; petioles present.
INFLORESCENCE: Axillary, flowers solitary; pedicles densely pubescent.
    CALYX: Sepals 5, bilabiate with 1 lobe bearing a prominent ridge.
    COROLLA: Petals 5, united, zygomorphic, bilabiate, blue with conspicuous blue and white spots on the lower lip; softly pubescent inside and outside.
    STAMENS: 4, epipetalous.
    PISTIL: Ovary superior; style 1, unbranched.
FRUIT: 4, bony nutlets in a persistent calyx.
COMMENTS: Frequent on various soils in prairies and openings in the Rio Grande Plains and Coastal Prairies. The leaves are occasionally eaten by white-tailed deer.

The closely-related *Scutellaria muriculata* Epl. has closely appressed stem pubescence and toothed leaf margins. [Not illustrated.]

LAMIACEAE (Labiatae)

DRUMMOND'S BETONY, PINK MINT
Stachys drummondii Benth.

ANNUAL: From a taproot.
STEMS: Erect, 4-angled, pubescent.
LEAVES: Simple, opposite, pubescent; blades ovate; margins toothed; apex obtuse; petioles pubescent.
INFLORESCENCE: Axillary with 2–3 flowers per node; pedicels pubescent.
    CALYX: Sepals 5, united about half their length; lobes acuminate, pubescent.
    COROLLA: Petals 5, united, zygomorphic, bilabiate, pink with lavender nectar guides against a white background; throat pubescent; at least 2 and one-half times longer than the calyx.
    STAMENS: 4, epipetalous.
    PISTIL: Ovary superior; style 1, slightly cleft above.
FRUIT: 4, bony nutlets.
COMMENTS: Frequent on various soils in chaparral, open woods, and brushy areas in the southern portion of the Rio Grande Plains and Coastal Prairies. The leaves are eaten by chachalacas and white-tailed deer.

LILIACEAE

1a. Tepals white with a purple, longitudinal midvein; bulbs lacking an onion-like odor. *Nothoscordum*
1b. Tepals white, pale-rose, or whitish-lavender, but lacking a conspicuously-colored midvein; bulbs with a strong onion-like odor. *Allium*

ELMENDORF ONION
Allium elmendorfii M. E. Jones ex Ownbey

PERENNIAL: From bulbs and usually with small, lateral bulblets; with a strong "onion" odor; bulbs not covered by a brown mesh-work of scales.
STEMS: A long, scabrous scape.
LEAVES: Simple, in a basal cluster; rounded to elliptic in cross section with a broad, longitudinal groove.
INFLORESCENCE: A terminal umbel subtended by 5–7 veined bracts; bracts glabrous, margins white-translucent.
    TEPALS: 6, free, white, pale-rose or whitish-lavender.
    STAMENS: 6, epipetalous, attached near the base of the inner tepals.
    PISTIL: Ovary superior; style 1, unbranched.
FRUIT: A capsule with 6 carpels.
COMMENTS: On sandy soils in the Rio Grande Plains and Coastal Prairies. The leaves and stems are eaten by white-tailed deer and cattle, and the bulbs are eaten by Rio Grande turkeys.

LILIACEAE

CROW-POISON
Nothoscordum bivalve (L.) Britt.

PERENNIAL: From a bulb, often with small bulblets but lacking an "onion-like" odor.
STEMS: A scape.
LEAVES: Simple, basal, glabrous; blades linear, succulent.
INFLORESCENCE: Umbellate, from a glabrous, elongated, naked scape; a scaly bract at base of umbel; pedicels glabrous.
    TEPALS: 6, free, white with a purple, longitudinal midvein.
    STAMENS: 6; anthers yellow; filaments widened at the base.
    PISTIL: Ovary superior, glabrous; style 1, unbranched.
FRUIT: A capsule.
COMMENTS: On sandy soils in a variety of habitats in the Rio Grande Plains and Coastal Prairies. The leaves and stems are eaten by white-tailed deer.

LINACEAE

1a. Inflorescence a raceme; stipular glands present. *Linum alatum*
1b. Inflorescence a panicle; stipular glands absent. *Linum rigidum*

FLAX
Linum alatum (Small) Winkler

ANNUAL or PERENNIAL
STEMS: Glabrous, slightly angled.
LEAVES: Simple, alternate, sessile, glabrous; blades linear; margins entire; stipular glands present.
INFLORESCENCE: A raceme.
    CALYX: Sepals 5, free, margins coarsely toothed and glandular; apex awned.
    COROLLA: Petals 5, free, yellow with reddish-orange veins below, actinomorphic, with a tuft of hairs at the insertion on the receptacle.
    STAMENS: 5, free; anthers yellow.
    PISTIL: Ovary superior, glabrous; style 1 with 5 branches near the apex.
FRUIT: A capsule separating into 5 segments.
COMMENTS: Frequent on sandy soils in the Rio Grande Plains and Coastal Prairies. The leaves and stems are eaten by white-tailed deer and cattle; the seeds are eaten by mourning doves.

LINACEAE

STIFFSTEM FLAX
Linum rigidum Pursh

ANNUAL: From a shallow taproot.
STEMS: Erect, glabrous.
LEAVES: Simple, alternate, glabrous; blades linear; margins entire; stipular glands absent
INFLORESCENCE: A panicle.
    CALYX: Sepals 5, free, lanceolate; midvein and lateral veins with minute, regularly spaced teeth; margins with stalked glands; apex awn-tipped.
    COROLLA: Petals 5, free, actinomorphic, rounded above; copper, orange to orange-yellow; readily dehiscent; with a tuft of hairs at the insertion on the receptacle.
    STAMENS: 5; filaments widened at the base and united into an inconspicuous ring; anthers yellow.
    PISTIL: Ovary superior; style 1 with 5 branches near the apex.
FRUIT: A capsule.
COMMENTS: Frequent on various soils in the Rio Grande Plains and Coastal Prairies. The leaves are eaten by white-tailed deer.

LOGANIACEAE

JUNIPERLEAF, POLLY-PRIM
Polypremum procumbens L.

ANNUAL or PERENNIAL
STEMS: Low-growing, freely branching, slightly ribbed, glabrous, turning orange with age.
LEAVES: Simple, opposite or whorled; blades linear to subulate; margins minutely ciliate, united at the base by a stipular line.
INFLORESCENCE: Flowers solitary in the leaf axils.
    CALYX: Sepals 4, united near the base, keeled; margins white-scarious.
    COROLLA: Petals 4, united below, white, actinomorphic.
    STAMENS: 4; anthers yellow.
    PISTIL: Ovary superior; style 1, unbranched.
FRUIT: A lobed, many-seeded capsule.
COMMENTS: Frequent on sandy soils in pastures, woods, and dunal areas in the Coastal Prairies and extreme eastern Rio Grande Plains. The leaves are eaten by white-tailed deer.

LYTHRACEAE

CALIFORNIA LOOSESTRIFE, HIERBA DEL CANCER
Lythrum californicum T.&G.

PERENNIAL
STEMS: Erect, angled or winged, glabrous.
LEAVES: Simple, alternate, sessile, glabrous; blades linear; margins entire; stipules present.
INFLORESCENCE: Flowers solitary in the leaf axils.
    CALYX: Sepals 5, united, reddish with subulate lobes.
    COROLLA: Petals 5, attached near the apex of the calyx tube, purple or rose-purple, about 6 mm long.
    STAMENS: 6, epipetalous, of different lengths.
    PISTIL: Ovary superior; style 1, unbranched; stigma capitate.
FRUIT: A cylindrical capsule.
COMMENTS: Frequent on clay soils in openings, pastures, and woodlands in the eastern portion of the Rio Grande Plains and Coastal Prairies. The leaves are eaten by white-tailed deer.

MALVACEAE

1a. Leaves palmately lobed. **2**
1b. Leaves entire or toothed, but not deeply lobed. **3**

2a. Flowers several near the stem apex; corollas orange-red. *Sphaeralcea*
2b. Flowers solitary at the nodes or on an elongated pedicel; corolla reddish-purple and white near the base. *Callirhoe*

3a. Fruit pulpy and berrylike; corolla erect, not spreading, crimson. *Malvaviscus*
3b. Fruit dry at maturity; corolla lobes usually spreading laterally. **4**

4a. Calyx not subtended by bracts. **5**
4b. Calyx subtended by bracts. **7**

5a. Calyx inflated in fruit and loosely enclosing the fruit. *Rhynchosida*
5b. Calyx not enclosing the fruit at maturity. **6**

6a. Leaves ovate to ovate-cordate; venation palmate. *Abutilon*
6b. Leaves usually linear or lanceolate; venation usually pinnate. *Sida*

7a. Shoots densely pubescent with white, feltlike, stellate pubescence. *Sphaeralcea*
7b. Shoots not as above. **8**

MALVACEAE

8a. Flowers in a terminal cluster. *Sida*
8b. Flowers usually solitary in the leaf axils. **9**

9a. Stems and sepals glabrous, but dotted with glands; fruit segments lacking bristles. *Cienfugosia*
9b. Stems and sepals pubescent, glands absent; fruit segments with bristles or spines. *Malvastrum*

PELOTAZO
Abutilon fruticosum Pen. & Rich.
[Syn. *Abutilon incanum* (Link) Sweet]

PERENNIAL or LONG-LIVED ANNUAL
STEMS: Slightly woody below; densely pubescent with pustulate-based hairs.
LEAVES: Simple, alternate, stellate-pubescent; blades ovate to ovate-cordate, venation palmate; margins toothed; apex obtuse; petioles present.
INFLORESCENCE: Flowers solitary in the leaf axils.
   CALYX: Sepals 5, united, triangular, recurved at the apex, about 3 mm long, including the tube.
   COROLLA: Petals 5, slightly united below, orange to copper-colored, red inside near the base; apex broadly rounded.
   STAMENS: Numerous (more than 10); monadelphous.
   PISTIL: Ovary superior; style branches numerous.
FRUIT: An elongated, 5-lobed capsule.
COMMENTS: Frequent on various soils in a diversity of habitats in the Rio Grande Plains. The leaves are eaten occasionally by white-tailed deer, and the seeds are eaten by bobwhite quail and mourning doves.

MALVACEAE

WINECUP, SLIMLOBE POPPYMALLOW
Callirhoe involucrata (Torr.) Gray var. *lineariloba* (T.&G.) Gray

PERENNIAL: From a radish-shaped root.
STEMS: Sprawling; pubescent with small, stellate hairs and longer straight hairs.
LEAVES: Simple, palmately lobed, alternate, pubescent; blade outline somewhat rounded; margins deeply toothed; petioles present; stipules auriculate.
INFLORESCENCE: Flowers solitary at the nodes, subtended by an involucel of linear, ciliate-margined bracts; pedicels pubescent, to 25 cm long.
    CALYX: Sepals 5, united near the base, lanceolate, with pustulate-based hairs and elevated veins, about 1/2 length of corolla.
    COROLLA: Petals 5, united near the base, erect, not spreading laterally, red to reddish-purple, white near the base, actinomorphic.
    STAMENS: Numerous, monadelphous.
    PISTIL: Ovary superior; styles numerous.
FRUIT: A capsule.
COMMENTS: Frequent on deep sands in prairies and openings in the Rio Grande Plains and Coastal Prairies. The leaves and flowers are eaten by white-tailed deer, and the leaves are eaten by cattle and Rio Grande turkeys.

YELLOW FUGOSIA, DRUMMOND FUGOSIA, SULPHUR MALLOW
Cienfuegosia drummondii (Gray) Lewton

PERENNIAL
STEMS: Glabrous, dotted with glands.
LEAVES: Simple, alternate, glabrous; blades ovate to rounded, venation palmate; margins wavy with some small, pointed teeth; petioles present; stipules linear, minute.
INFLORESCENCE: Flowers solitary from the leaf axils on an elongated, glabrous pedicel; involucel present below the calyx; bracts mostly spatulate, some with a soft bristle at the apex.
    CALYX: Sepals 5, free, glandular-dotted, acute, obtuse or with a minute spine at the apex.
    COROLLA: Petals 5, united near the base, actinomorphic, sulphur-yellow, 2.5–3.5 cm long.
    STAMENS: Numerous; monadelphous; anthers yellow.
    PISTIL: Ovary superior; style 1 with 5 velvet-red stigmas.
FRUIT: A capsule with large, slightly pubescent seeds.
COMMENTS: Frequent on clays, usually in swales or other poorly drained places in the Coastal Prairies. The leaves are eaten by white-tailed deer and cattle. The fruits are an alternate host for boll weevils.

MALVACEAE

Malvastrum Gray
1a. Carpels horseshoe-shaped, lateral spines ciliate; shoots appressed-pubescent. ***Malvastrum coromandelianum***
1b. Carpels not as above, lateral spines not ciliate; shoots stellate pubescent. ***Malvastrum aurantiacum***

WRIGHT FALSE MALLOW
Malvastrum aurantiacum (Scheele) Walp.

PERENNIAL
STEMS: Erect or trailing; stellate-pubescent.
LEAVES: Simple, alternate, stellate-pubescent; blades ovate; margins dentate; petioles and stipules present.
INFLORESCENCE: Flowers solitary in the leaf axils.
   CALYX: Subtended by an involucel of 3 bracts; sepals 5, united below, lobes broadly triangular, stellate-pubescent.
   COROLLA: Petals 5, united near the base, actinomorphic, yellow-orange; apex irregularly lobed with a tuft of hairs at base, 1.5–2.0 cm long.
   STAMENS: Numerous; monadelphous; anthers yellow.
   PISTIL: Ovary superior, pubescent; style branched.
FRUIT: A capsule with up to 12 segments, each with stiff, horizontal bristles, each carpel with a single, large, elliptical seed.
COMMENTS: Frequent on clay soils in openings and along roads in the Coastal Prairies. The leaves are eaten by white-tailed deer and cattle.

MALVACEAE

THREELOBE FALSE MALLOW
Malvastrum coromandelianum (L.) Gke.

PERENNIAL
STEMS: Erect or sprawling; appressed-pubescent.
LEAVES: Simple, alternate, pubescent; blades ovate to ovate-lanceolate; margins crenate; apex obtuse; petioles pubescent with white hairs; stipules linear.
INFLORESCENCE: Flowers solitary in the leaf axils.
    CALYX: Subtended by an involucel of 3 linear bracts; sepals 5, united for about one-third their length, triangular, pubescent.
    COROLLA: Petals 5, slightly united near the base; yellow, usually opening in the afternoon on sunny days.
    STAMENS: Numerous; monadelphous.
    PISTIL: Ovary superior, pubescent.
FRUIT: A pubescent capsule separating into numerous 1-seeded carpels; carpels "horseshoe-shaped," with white, lateral bristles and 1 prominent, brown, lateral spine, the spine with white cilia.
COMMENTS: Frequent on loam and clay soils in openings, stream bottoms, and roadsides in the Rio Grande Plains and Coastal Prairies. The leaves and stems are eaten by white-tailed deer and cattle.

TEXAS MALLOW
Malvaviscus arboreus Cav. var. *drummondii* (T.&G.) Schery.
[Syn. *Malvaviscus drummondii* T.&G.]

PERENNIAL: Often shrubby.
STEMS: Erect or scandent; minutely stellate-pubescent.
LEAVES: Simple, alternate, stellate-pubescent; blades broadly-ovate, venation palmate, cordate at the base; margins broadly-crenate; apex obtuse; petioles present.
INFLORESCENCE: Flowers solitary from the leaf axils.
    CALYX: Subtended by an involucel of linear, pubescent bracts; sepals 5, united, with broadly triangular lobes.
    COROLLA: Petals 5, erect, united near the base, crimson.
    STAMENS: Numerous; monadelphous.
    PISTIL: Ovary superior; style 1, red, with 10 linear branches near the apex.
FRUIT: Red, pulpy, berrylike.
COMMENTS: Frequent on loamy soils in pastures and woods in the eastern portion of the Rio Grande Plains and Coastal Prairies.
The leaves are eaten by white-tailed deer.

MALVACEAE

SPEARLEAF SIDA
Rhynchosida physocalyx (Gray) Fryx.
[Syn. *Sida physocalyx* Gray]

PERENNIAL
STEMS: Usually trailing or reclining; stellate-pubescent.
LEAVES: Simple, alternate, stellate-pubescent; blades broadly ovate, venation palmate; margins toothed; apex obtuse; petioles elongated, stellate-pubescent; stipules minute.
INFLORESCENCE: Flowers solitary from the leaf axils; pedicels densely stellate-pubescent.
    CALYX: Sepals 5, united below, pubescent; inflated and persistent in fruit.
    COROLLA: Petals 5, united below, light yellow, actinomorphic.
    STAMENS: Numerous; monadelphous.
    PISTIL: Ovary superior, with 10 lobes.
FRUIT: A lobed capsule.
COMMENTS: Frequent on sandy and clay loam soils in prairies, openings and waste places in the Rio Grande Plains and Coastal Prairies. The leaves are eaten by white-tailed deer and cattle.

MALVACEAE

Sida L.

1a. Flowers in a terminal cluster; calyx subtended by linear bracts; corollas red, orange or salmon-colored. *Sida ciliaris*

1b. Flowers 1-several from the leaf axils; calyx not subtended by bracts; corolla yellow or yellow-orange. **2**

2a. Leaves with a blunt spine near the base of the petiole. *Sida spinosa*

2b. Leaves lacking a blunt spine near the base of the petiole. **3**

3a. Plants mostly prostrate or reclining; leaves broadly lanceolate. *Sida abutifolia*

3b. Plants mostly erect and branching above; leaves linear. *Sida lindheimeri*

SPREADING SIDA
Sida abutifolia Mill.
[Syn. *Sida filicaulis* T.&G.]

PERENNIAL
STEMS: Mostly prostrate or reclining; stellate-pubescent.
LEAVES: Simple, alternate, stellate-pubescent; blades broadly lanceolate with a cordate base; margins crenate; apex obtuse; petioles stellate-pubescent; stipules present.
INFLORESCENCE: 1-several flowers per node; pedicel usually less than length of blade.
   CALYX: Sepals 5, united near the base; lobes deltoid; apex acute to acuminate; stellate-pubescent.
   COROLLA: Petals 5, united below, actinomorphic, golden yellow, 6–8 mm long.
   STAMENS: Numerous; monadelphous; anthers yellow.
   PISTIL: Ovary superior, yellow; style with several branches.
FRUIT: A capsule.
COMMENTS: Frequent on sand, clay, or caliche in open woodlands in the Rio Grande Plains and Coastal Prairies. The leaves are eaten by white-tailed deer and cattle.

MALVACEAE

BRACTED SIDA
Sida ciliaris L. var. *mexicana* (Moric.) Shinners

PERENNIAL: From a woody rootstock.
STEMS: Pubescent with reflexed hairs.
LEAVES: Simple, alternate; glabrous above, stellate-pubescent below; blades linear to lanceolate and cordate at base; margins toothed mostly near the apex; petioles grooved; stipules linear.
INFLORESCENCE: Flowers in a terminal cluster.
    CALYX: Subtended by linear, pubescent bracts; sepals 5, united; lobes triangular, lobed, pubescent, persistent in fruit.
    COROLLA: Petals 5, slightly united near the base, red, orange or salmon-colored.
    STAMENS: Numerous; monadelphous.
    PISTIL: Ovary superior; styles several.
FRUIT: A glabrous capsule.
COMMENTS: Frequent on sand or clayey loams in prairies, openings, and waste places in the Rio Grande Plains and Coastal Prairies. The leaves and stems are eaten by cattle.

MALVACEAE

PRICKLY SIDA, PRICKLY MALLOW
Sida spinosa L.

ANNUAL: From a taproot.
STEMS: Erect, branching above; stellate-pubescent with hairs of 2 different sizes.
LEAVES: Simple, alternate, pubescence similar to stems; blades linear-lanceolate to lanceolate, 2–5 cm long; margins crenate; apex obtuse; petioles with a small, blunt spine near the base; stipules present.
INFLORESCENCE: Flowers 1-several arising from the leaf axils.
    CALYX: Sepals 5, united near the base, margins often purple-tinged, pubescent, persisting around the fruit.
    COROLLA: Petals 5, united near the base, actinomorphic, yellow, 3–9 mm long.
    STAMENS: Numerous; monadelphous; anthers yellow.
    PISTIL: Ovary superior; styles numerous.
FRUIT: A capsule with spine-tipped carpel segments.
COMMENTS: Frequent on various soils in prairies, openings, and waste places in the Rio Grande Plains and Coastal Prairies. The leaves and stems are eaten by white-tailed deer and cattle.

SHOWY SIDA
Sida lindheimeri Engelm. & Gray

Similar to *S. spinosa,* but lacking the characteristic blunt spine near the base of the petiole. The species is typically a more robust, glabrous perennial that occurs on sandy soils in woodlands and brushy areas in the Rio Grande Plains and Coastal Prairies. The leaves are eaten by white-tailed deer and cattle. [Not illustrated.]

MALVACEAE

Sphaeralcea St. Hil.

1a. Shoots densely pubescent with white, stellate hairs; leaves ovate to cordate. ***Sphaeralcea lindheimeri***

1b. Shoots not as above; blades palmately divided. ***Sphaeralcea pedatifida***

WOOLLY GLOBEMALLOW
Sphaeralcea lindheimeri Gray

PERENNIAL
STEMS: Sprawling; all shoots densely pubescent with white, feltlike, stellate hairs.
LEAVES: Simple, alternate, densely pubescent; blades broadly ovate to cordate, venation palmate; margins crenate; apex obtuse; stipules linear, pubescent.
INFLORESCENCE: Flowers in several-flowered, terminal clusters or solitary in the leaf axils.
  CALYX: Subtended by an involucel of linear, pubescent bracts; sepals 5, united, densely pubescent, nearly glabrous within.
  COROLLA: Petals 5, united at the base, orange.
  STAMENS: Numerous; monadelphous.
  PISTIL: Ovary superior; style branches numerous; stigmas red.
FRUIT: A capsule.
COMMENTS: Frequent on sandy soils in openings and prairies in the southern portion of the Rio Grande Plains and Coastal Prairies. The leaves are eaten by white-tailed deer and cattle.

PALMLEAF GLOBEMALLOW
Sphaeralcea pedatifida Gray

PERENNIAL: From a woody rootstock.
STEMS: Sprawling, branching at the nodes; stellate-pubescent.
LEAVES: Simple, palmately divided, alternate, stellate-pubescent; segments deeply lobed; petioles present.
INFLORESCENCE: Flowers several near the stem apex.
  CALYX: Sepals 5, united, lobes acuminate at the apex, stellate-pubescent.
  COROLLA: Petals 5, slightly united near the base, orange-red, about 1 cm long.
  STAMENS: Numerous; monadelphous.
  PISTIL: Ovary superior; style branches several.
FRUIT: A capsule.
COMMENTS: Frequent on sandy soils in openings and brushlands in the Rio Grande Plains. The leaves are eaten by white-tailed deer.

MARSILEACEAE

LARGE-FOOT PEPPERWORT, WATER-CLOVER
Marsilea macropoda A.Br.

PERENNIAL
STEMS: With wiry rhizomes; aerial stems absent; forming colonies.
LEAVES: Compound; leaflets 4 (resembles a four-leaf clover), pubescent; leaflets obtuse at the base; apex rounded; often emergent from shallow, standing water.
SPOROCARP: 2–6 in a cluster, nut-like, subterranean.
COMMENTS: Frequent in swales, ditches, and other low damp places in the Rio Grande Plains and Coastal Prairies. The leaves and stems are eaten by white-tailed deer, javelinas, feral pigs, and cattle, while bobwhite quail consume the leaves.

MENISPERMACEAE

ORIENT-VINE, CORREHUELA
Cocculus diversifolius DC.

PERENNIAL: Plants dioecious.
STEMS: A twining vine; becoming woody.
LEAVES: Simple, alternate, glabrous; blades variously shaped; vein pattern branching from the base of the blade; margins entire; apex often with a mucronate tip; petioles on new growth minutely pubescent.
INFLORESCENCE: Flowers in small panicles from the upper leaf axils.
STAMINATE FLOWERS
    CALYX: Subtended by irregularly shaped, whitish bracts; sepals 3, free, creamy white; apex rounded, much longer than petals.
    COROLLA: Petals 6, free, cream-colored.
    STAMENS: 6, free; anthers pointed toward the midsection of the flower (resembles a snake ready to strike).
PISTILLATE FLOWERS: (Perianth segments similar to the staminate flowers)
    PISTIL: Ovary superior.
FRUIT: A small, fleshy, purple-black drupe.
COMMENTS: Frequent on sandy soils in pastures and openings in the Rio Grande Plains and Coastal Prairies. The leaves are eaten by white-tailed deer.

NYCTAGINACEAE

BERLANDIER TRUMPETS
Acleisanthes obtusa (Choisy) Standl.

PERENNIAL
STEMS: Vinelike; glabrous to minutely pubescent.
LEAVES: Simple, opposite, glabrous; blades broadly ovate and deltoid-truncate at the base; apex broadly rounded; petioles present.
INFLORESCENCE: Flowers usually 3 arising from the upper nodes, subtended by linear bracts.
    CALYX: Petaloid, tubular, reddish on the tube, lobes white with prominent purple keels on the outer surface; 3.5–5.5 m long, tube minutely pubescent.
    COROLLA: Absent.
    STAMENS: 5, exserted; filaments purple-pink.
    PISTIL: Ovary superior; style 1, unbranched.
FRUIT: A narrowly oblong anthocarp.
COMMENTS: Frequent in brushy pastures on sand, clay, and caliche soils in the Rio Grande Plains and Coastal Prairies. The leaves are eaten by white-tailed deer.

NYMPHAEACEAE

SENORITA WATERLILY, BLUE WATER LILY, LAMPAZOS
Nymphaea elegans Hook.

PERENNIAL
STEMS: Short rhizomes present; aerial stems absent.
LEAVES: Simple, arising from the rhizome, subpeltate, floating, glabrous; blades rounded or elliptic; margins wavy; upper epidermis dark green, the lower with prominent, elevated, purplish veins; petioles attached adjacent to a V-shaped cleft, submersed.
INFLORESCENCE: Flowers solitary on an elongated pedicel.
    CALYX and COROLLA: Present, but best termed as tepals; in bud stage with 4, free sepals and numerous (more than 10) free petals, actinomorphic.
    STAMENS: Numerous (more than 10), attached to a hypanthium; anthers and filaments not clearly differentiated.
    PISTIL: Ovary superior; style radiate.
FRUIT: A capsule with numerous seeds.
COMMENTS: Frequent in ponds, lakes, streams, swales, and ditches in the Rio Grande Plains and Coastal Prairies. The leaves and stems are eaten by sandhill cranes and several species of waterfowl.

OLEACEAE

LOW MENODORA
Menodora heterophylla Moric.

PERENNIAL
STEMS: Low-growing, slightly woody at the base, minutely pubescent.
LEAVES: Simple, opposite, nearly glabrous; blades lobed or entire; margins minutely ciliate.
INFLORESCENCE: Flowers solitary at the apex.
    CALYX: Sepals 10, united, lobes linear, slightly scabrous.
    COROLLA: Petals 5 (6), united, yellow within, reddish on the back, slightly zygomorphic, pubescent in the throat.
    STAMENS: 2, epipetalous.
    PISTIL: Ovary superior; style 1, unbranched; stigma capitate.
FRUIT: A capsule with 4 seeds.
COMMENTS: Frequent on clays, heavier loams, caliche outcrops, or hills in the Rio Grande Plains and western portion of the Coastal Prairies. The leaves are eaten by white-tailed deer and bobwhite quail.

ONAGRACEAE

1a. Petals narrowed to the base; inflorescence spicate or a remotely flowered spike, slightly zygomorphic. *Gaura*
1b. Petals usually rounded at the base, actinormorphic; inflorescence few-flowered from the leaf axils. *Oenothera*

PLAINS GAURA
Gaura brachycarpa Small
[Syn. *Gaura tripetala* Cav. var. *coryi* Munz]

ANNUAL: From a taproot.
STEMS: Sprawling and arching upward near the apex; pubescent with white, nearly straight, perpendicular hairs.
LEAVES: Simple, alternate, pubescent; blades lanceolate to narrowly lanceolate; margins entire to wavy or sinuate.
INFLORESCENCE: Spicate or a remotely flowered spike; flowers subtended by small bracts.
   CALYX: Sepals 4, attached to an elongated hypanthium, free, pubescent on the back.
   COROLLA: Petals 4, slightly zygomorphic, free, rose, lavender or red; clawed at the base.
   STAMENS: 8; filaments white, slightly flattened; anthers crimson.
   PISTIL: Ovary inferior, 4-lobed; style 1.
FRUIT: A sessile, ovoid or pyramid-shaped, 4-ribbed capsule; pubescent with hairs oriented downward, about 7 mm long.
COMMENTS: Frequent on sand in the Rio Grande Plains and Coastal Prairies. The leaves are eaten by white-tailed deer and cattle. The seeds are eaten by mourning doves, bobwhite quail, scaled quail, and Rio Grande turkeys.

ONAGRACEAE

Oenothera
1a. Corolla pink or rose. *Oenothera speciosa*
1a. Corolla yellow. **2**

2a. Petals 2.0–3.5 cm long. *Oenothera grandis*
2b. Petals 0.5–1.8 cm long. *Oenothera laciniata*

EVENING PRIMROSE
Oenothera grandis (Britt.) Smyth

ANNUAL: From a taproot.
STEMS: Low-growing, densely pubescent.
LEAVES: Simple, alternate, pubescent; blades lanceolate, oblanceolate, lyrate; margins lobed nearly to the midrib.
INFLORESCENCE: Flowers several in the leaf axils.
    CALYX: Sepals 4, turning down when corolla opens, attached at apex of hypanthium, pubescent, 2.0–2.5 cm long; hypanthium up to 4 cm long.
    COROLLA: Petals 4, free, actinomorphic, lemon-yellow, 2.0–3.5 cm long.
    STAMENS: 8; filaments attached at midpoint of anthers.
    PISTIL: Ovary inferior, densely pubescent; style 1, branches 4.
FRUIT: A capsule.
COMMENTS: Frequent on sandy soils in the Coastal Prairies and Rio Grande Plains. The leaves are eaten by white-tailed deer and cattle, and the seeds are consumed by mourning doves and white-winged doves.

DOWNY EVENING PRIMROSE, CUTLEAF EVENING PRIMROSE
Oenothera laciniata Hill.

Similar to *O. grandis* except it has sepals 0.5–1.3 cm long and petals 0.5–1.8 cm long. [Not illustrated.]

ONAGRACEAE

MEXICAN EVENING PRIMROSE, AMAPOLA DEL CAMPO
Oenothera speciosa Nutt.

PERENNIAL
STEMS: Densely pubescent with straight hairs.
LEAVES: Simple, alternate, pubescent, some hairs up to 1 mm long; blades oblong or oblanceolate, pinnatisect.
INFLORESCENCE: Several flowers from the leaf axils; flowers drooping in larger plants.
  CALYX: Sepals 4, united, pubescent, tearing longitudinally along vertical red lines where sepals are united, attached at apex of a hypanthium; hypanthium up to 1.5 cm long.
  COROLLA: Petals 4, free, pink or rose, actinomorphic, 2.5–3.0 cm long; nectar guides prominent with a yellow-green area at base.
  PISTIL: Ovary inferior; style 1, branches 4.
FRUIT: A pubescent, clublike capsule.
COMMENTS: Frequent on clay or loamy soils in prairies, openings, fields, and along roads in the Rio Grande Plains and Coastal Prairies. The leaves are eaten by white-tailed deer and Rio Grande turkeys.

OXALIDACEAE

1a. Corolla purple-violet; a bulb present. *Oxalis drummondii*
1b. Corolla yellow; bulb absent. *Oxalis dillenii*

DILLENS OXALIS, WOODSORREL, SOUR CLOVER
Oxalis dillenii Jacq.

ANNUAL
STEMS: Low-growing, creeping; pubescent.
LEAVES: Palmately trifoliolate, alternate, pubescent; leaflets acute or obcordate at the base; margins entire; apex rounded and cleft; petioles pubescent; stipules scalelike, pubescent.
INFLORESCENCE: Racemose or paniculate; pedicels pubescent.
   CALYX: Sepals 5, free, pubescent, lanceolate; apex obtuse; about 1/2 length of the corolla.
   COROLLA: Petals 5, loosely united or free at the base, yellow; rounded at the apex.
   STAMENS: 10; monadelphous; in 2 series of 5 stamens each; shorter stamens two-thirds to three-quarters the length of the longer.
   PISTIL: Ovary superior; styles 5.
FRUIT: An elongated capsule with numerous seeds.
COMMENTS: Frequent on sandy, rocky or gravelly soils in the Rio Grande Plains and Coastal Prairies. The leaves and stems are an important food of white-tailed deer and are also eaten by Rio Grande turkeys and cattle.

WOOD SORREL, DRUMMOND OXALIS
Oxalis drummondii Gray

PERENNIAL: From a small bulb.
STEMS: Aerial stems absent.
LEAVES: Trifoliolate, basal (arising from the bulb), glabrous; leaflets in the shape of an inverted equilateral triangle; margins entire with scattered hairs; apex slightly cleft.
INFLORESCENCE: A few-flowered umbellate cluster; peduncles and pedicels glabrous.
   CALYX: Sepals 5, free, with a brown patch on the outer surface near the apex.
   COROLLA: Petals 5, united near base, actinomorphic, purple-violet.
   STAMENS: 10, united near base.
   PISTIL: Ovary superior; styles 3–5.
FRUIT: A capsule with round seeds.
COMMENTS: Common on sandy-gravelly soils in pastures, woods, stream bottoms, and waste places in the Rio Grande Plains and Coastal Prairies. The leaves are eaten by sandhill cranes.

PAPAVERACEAE

SPINY PRICKLEPOPPY, RED POPPY
Argemone sanguinea Greene

ANNUAL: From a taproot; shoots with a viscous, yellow latex.
STEMS: Erect, bluish-green with stiff, minutely pubescent spines.
LEAVES: Simple, alternate, bluish-green, spiny; blades pinnatisect, white on the upper epidermis over the primary veins; margins spine-tipped; petioles present below, subsessile above.
INFLORESCENCE: A terminal panicle with several spiny bracts subtending the flowers.
    CALYX: Sepals 2–3, with spiny, hornlike crests, not persisting.
    COROLLA: Petals 5, free, often wrinkled, white, actinomorphic, rounded at the apex, 3.5–4.0 cm long.
    STAMENS: Numerous (many more than 10); filaments and anthers golden.
    PISTIL: Ovary superior; style 1, with 4 stigmatic surfaces (resembles a starfish).
FRUIT: A spiny capsule; seeds brown with a warty seed coat.
COMMENTS: Frequent on loam and sandy soils in disturbed areas in the Rio Grande Plains and Coastal Prairies. The seeds are eaten by bobwhite quail and mourning doves.

PHYTOLACCACEAE

BLOODBERRY, ROUGEPLANT, PIGEONBERRY, CORALITO
Rivina humilis L.

PERENNIAL: From a slightly woody base.
STEMS: Erect, branching at the nodes, glabrous.
LEAVES: Simple, alternate, glabrous; blades ovate to ovate-lanceolate; margins sinuate; apex obtuse.
INFLORESCENCE: A terminal or occasionally axillary raceme; pedicels pink or rose; with linear bracts subtending the pedicels.
    CALYX: Sepals 4, free, actinomorphic, white, rose, or reddish.
    COROLLA: Petals absent.
    STAMENS: 4; filaments and anthers white.
    PISTIL: Ovary superior; style 1, unbranched.
FRUIT: A small red berry.
COMMENTS: On various soils in brushy pastures, mottes, and bottom woods (usually in shade) in the Rio Grande Plains and Coastal Prairies. The fruits are eaten by Rio Grande turkeys, chachalacas, mourning doves, white-winged doves and numerous species of passerine birds. The leaves are consumed by javelinas.

PLANTAGINACEAE

1a. Leaves linear; margins mostly entire. *Plantago hookeriana*
1b. Leaves obovate to oblanceolate; margins remotely toothed. *Plantago rhodosperma*

TALLOW WEED, HOOKER'S PLANTAIN
Plantago hookeriana Fisch. & Mey.

ANNUAL: From a taproot.
STEMS: An acaulescent scape, pubescent.
LEAVES: Simple, in a basal cluster, appressed-pubescent; blades linear, grayish-green; margins mostly entire.
INFLORESCENCE: A terminal spike with bracts subtending each flower; bracts with green centers and translucent margins, shorter than the calyx.
    CALYX: Sepals 4, free, with broad, scarious margins.
    COROLLA: Petals 4, connate, lobes translucent with lavender spots near the throat, spreading laterally.
    STAMENS: 4, epipetalous; filaments elongated.
    PISTIL: Ovary superior, pubescent; style 1.
FRUIT: A capsule.
COMMENTS: Frequent on sandy soils in prairies, openings, and waste places in the Rio Grande Plains and Coastal Prairies. The leaves are eaten by Rio Grande turkeys, white-tailed deer, and cattle. The seeds are consumed by bobwhite quail and mourning doves.

REDSEED PLANTAIN
Plantago rhodosperma Dcne.

ANNUAL: From a shallow taproot.
STEMS: Acaulescent.
LEAVES: Simple, in a basal rosette, pubescent; blades obovate to oblanceolate; margins remotely toothed; petioles present.
INFLORESCENCE: A spike on a pubescent scape; flowers subtended by pubescent bracts; bracts acute at the apex.
    CALYX: Sepals 4, free, pubescent with broad, scarious margins.
    COROLLA: Petals 4, connate, brownish or scarious, with a slightly elevated rib on the back.
    STAMENS: 4, epipetalous.
    PISTIL: Ovary superior, style pubescent.
FRUIT: A capsule.
COMMENTS: Common on clay or heavier sands in prairies and openings in the Rio Grande Plains and Coastal Prairies. The leaves are eaten by white-tailed deer and cattle, and the seeds are eaten by bobwhite quail and mourning doves. This plant is also eaten by the Texas tortoise.

POLEMONIACEAE

PHLOX
Phlox drummondii Hook.

ANNUAL: From a taproot.
STEMS: Densely pubescent.
LEAVES: Simple, opposite below, alternate above, sessile, pubescent; blades narrowly lanceolate; margins entire.
INFLORESCENCE: A raceme with leafy bracts.
  CALYX: Sepals 5, united, lobes acuminate, pubescent, less than one-half the length of the corolla.
  COROLLA: Petals 5, united, purple-lavender; tube densely pubescent on the outside with a dark purple, star-shaped entrance at the apex; upper zone of the throat white, the lower deep purple; lobes extended laterally, broadly rounded.
  STAMENS: 5, epipetalous, 3 long and 2 short, included within the throat of the corolla; anthers yellow.
  PISTIL: Ovary superior; style 1, branches 3; included within the throat of the corolla.
FRUIT: A capsule.
COMMENTS: Common on sandy soils in openings and prairies in the Rio Grande Plains and Coastal Prairies. The leaves are eaten by white-tailed deer and bobwhite quail.

GOLDSMITH PHLOX
Phlox drummondii Hook. ssp. *wilcoxiana* (Bogusch) Wherry
[Syn. *Phlox drummondii* Hook. var. *wilcoxiana* (Bogusch) Whiteh.]
[Syn. *Phlox goldsmithii* Whiteh.] This subspecies has a deep red corolla.
[Not illustrated.]

POLYGONACEAE

1a. Stems tomentose; leaves sessile; stipular sheath absent. *Eriogonum*
1b. Stems glabrous; leaves with petioles; stipular sheath (ocrea) present. **2**

2a. Sepals 6, the inner inflated and winged. *Rumex*
2b. Sepals 5, not inflated nor winged. *Polygonum*

HEART-SEPAL WILD BUCKWHEAT
Eriogonum multiflorum Benth.

ANNUAL: From a taproot.
STEMS: Erect, tomentose, with white, matted hairs.
LEAVES: Simple, alternate, sessile, tomentose; blades ovate, lower epidermis white; margins wavy, revolute; apex obtuse.
INFLORESCENCE: Paniculate or umbellate, arising from a cuplike structure consisting of 5 pubescent bracts.
  CALYX: Sepals 6, winged, stipitate, white.
  COROLLA: Petals absent.
  STAMENS: 9, free.
  PISTIL: Ovary superior.
FRUIT: An achene.
COMMENTS: Frequent and locally abundant on sandy soils in the Rio Grande Plains and Coastal Prairies. The leaves are occasionally eaten by white-tailed deer, and the seeds are eaten by mourning doves and Rio Grande turkeys.

POLYGONACEAE

PENNSYLVANIA SMARTWEED, PINK SMARTWEED
Polygonum pensylvanicum L.
[Syn. *Persicaria pensylvanica* (L.) Small] [Syn. *Polygonum bicorne* (Raf.)
Nieuw.] [Syn. *Persicaria bicornis* Raf.]

ANNUAL: Long-lived.
STEMS: Red at the nodes, glabrous.
LEAVES: Simple, alternate, glabrous; blades lanceolate; margins entire,
minutely ciliate; petioles red at the base; stipules forming a scaly sheath
(ocrea).
INFLORESCENCE: A densely clustered spicate raceme; bracts present in
the axis; peduncle with dense, stalked, red-tipped, glandular hairs.
  CALYX: Sepals 5, slightly united at the base, rose-pink, apex rounded,
  about 4 mm long.
  COROLLA: Petals absent.
  STAMENS: 8, attached at the base of the petaloid calyx.
  PISTIL: Ovary superior, 3 angled; style 1, branches 3.
FRUIT: An achene.
COMMENTS: Frequent in wet areas, ditches, and disturbed areas in the Rio
Grande Plains and Coastal Prairies. The leaves are eaten by white-tailed
deer, and the seeds are consumed by mourning doves and Rio Grande
turkeys.

DOCK, AMNASTLA
Rumex chrysocarpus Moris

ANNUAL: From a taproot.
STEMS: Erect, often branching from the lower nodes, glabrous.
LEAVES: Simple, alternate, glabrous; blades lanceolate with a conspicuous
midvein; margins crisped or occasionally entire; petioles present; stipules
membranous, scalelike (ocrea).
INFLORESCENCE: In axillary clusters over much of the stem.
  CALYX: Sepals 6, the outer 3 inconspicuous and linear, the inner 3
  inflated with lateral spines; wings extending from a centrally located,
  white, grainy tubercle.
  COROLLA: Petals absent.
  STAMENS: 6.
  PISTIL: Ovary superior; styles 3.
FRUIT: A trigonous achene.
COMMENTS: Common on clay or heavier sands in ditches, swales, and
stream bottoms in the Rio Grande Plains and Coastal Prairies. The leaves
are occasionally eaten by white-tailed deer, while the seeds are eaten by
Rio Grande turkeys.

PORTULACACEAE

1a. Corolla yellow; leaves opposite, blades mostly ovate. *Portulaca oleracea*
1b. Corolla purple or reddish-purple; leaves alternate, blades lanceolate.
Portulaca pilosa

COMMON PURSLANE, VERDOLAGA
Portulaca oleracea L.

ANNUAL: From a taproot.
STEMS: Erect or prostrate; not rooting at the nodes; succulent, glabrous.
LEAVES: Simple, opposite or whorled near the apex, occasionally alternate; blades ovate, succulent; margins entire; apex rounded; petioles present.
INFLORESCENCE: Flowers in terminal clusters.
    CALYX: Sepals 2, cleft (resembles the Pope's hat), attached to a short hypanthium.
    COROLLA: Petals 5, slightly united near the base, yellow, attached to a short hypanthium; apex cleft.
    STAMENS: 10, free; anthers yellow.
    PISTIL: Ovary partially-inferior; perigynous.
FRUIT: A many-seeded capsule.
COMMENTS: Common on various soils in fields, disturbed areas and waste places in the Rio Grande Plains and Coastal Prairies. The leaves are eaten by white-tailed deer and cattle.

CHISME, SHAGGY PORTULACA
Portulaca pilosa L.
[Syn. *Portulaca mundula* I. M. Johnston]

ANNUAL: From a branched taproot.
STEMS: Prostrate, succulent, with white, matted hairs.
LEAVES: Simple, alternate, succulent; blades lanceolate; margins entire; apex obtuse; petioles present.
INFLORESCENCE: Flowers 1-several from the leaf axils.
    CALYX: Sepals 2 or 2 cleft, margins translucent with a green awn near the midsection.
    COROLLA: Petals 5, purple, slightly united at the base, apex rounded.
    STAMENS: 10–15; adnate near the base; anthers yellow.
    PISTIL: Ovary inferior; style 1, with 5 spreading, purple branches.
FRUIT: A many-seeded capsule.
COMMENTS: Common on sand, loam, and gravelly soils in the Rio Grande Plains and western portion of Coastal Prairies. The leaves are an important food of white-tailed deer, javelinas and feral pigs, and both the leaves and seeds are eaten by scaled quail.

RANUNCULACEAE

1a. Plants vines with twining petiolules; sepals 4. *Clematis*
1b. Plants erect herbs, not twining; sepals as many as 20. *Anemone*

Anemone L.
1a. Flowering stems glabrous below the involucre, densely pubescent above. *Anemone caroliniana*
1b. Flowering stems densely pubescent over their length. *Anemone berlandieri*

CAROLINA ANEMONE
Anemone caroliniana Walt.

PERENNIAL: From a tuberous base and with slender rhizomes.
STEMS: Flowering stems glabrous below the involucre, densely pubescent above.
LEAVES: Compound with 3 segments, each 2–3-lobed, basal, glabrous; lobes toothed; petioles present; involucre present below the midpoint of the flowering axis, whorled; segments linear or lanceolate.
INFLORESCENCE: Flowers solitary on elongated peduncles.
   CALYX: Sepals 10–20, free, white, lavender, or blue, actinomorphic, appressed pubescent on the outer sepals.
   COROLLA: Petals absent; calyx petaloid.
   STAMENS: Numerous, more than 10.
   PISTIL: Pistils numerous, simple; ovary superior.
FRUIT: An achene included in a mass of white, woolly hairs.
COMMENTS: Infrequent in sandy prairies and openings in the Coastal Prairies. The leaves are occasionally eaten by white-tailed deer.

TENPETAL ANEMONE
Anemone berlandieri Pritz.
[Syn. *Anemone heterophylla* Nutt.] [Syn. *Anemone decapetala* Ard. var. *heterophylla* (Nutt.) Britt.]
Similar to *A. caroliniana*. However, the stems are pubescent throughout. This species is frequent on sand or clay soils in openings and prairies in the northern portion of the Rio Grande Plains and Coastal Prairies. The leaves are eaten by white-tailed deer. [Not illustrated.]

RANUNCULACEAE

TEXAS VIRGIN'S-BOWER, BARBAS DE CHIVATO
Clematis drummondii T.&G.

PERENNIAL: Usually monoecious or dioecious.
STEMS: A ribbed, densely pubescent vine.
LEAVES: Compound with 5–7 variously lobed and dissected leaflets, opposite, pubescent; uppermost leaves simple; petiolules twining.
INFLORESCENCE: Flowers solitary from the leaf axils on elongated, pubescent pedicels.
   CALYX: Sepals 4, free, yellowish-green, softly pubescent, readily dehiscent.
   COROLLA: Petals absent.
   STAMENS: Numerous, more than 10 per flower; filaments green, flattened; anthers not much wider than the filaments.
   PISTIL: Pistils numerous per flower; styles unbranched.
FRUIT: Achenes numerous per pistillate flower; styles plumose, persistent.
COMMENTS: Common on sandy loam and clay loam or caliche soils in pastures and woodlands in the Rio Grande Plains and Coastal Prairies. The leaves are an important food of white-tailed deer; Rio Grande turkeys and cattle also occasionally eat the leaves.

RUBIACEAE

1a. Plants annuals; leaves linear to narrowly lanceolate. *Diodia*
1b. Plants perennials; leaves ovate, broadly rounded or lanceolate.
   Richardia

ROUGH BUTTONWEED, POOR-JOE
Diodia teres Walt.

ANNUAL: From a taproot.
STEMS: Branching above the base, often purple-tinged, pubescent with hairs of 2 different lengths.
LEAVES: Simple, opposite, scabrous; blades linear to narrowly lanceolate; margins with minute teeth; apex acute; stipular bristles numerous, the larger pink to purple.
INFLORESCENCE: Flowers 1–3 in the leaf axils.
   CALYX: Sepals 4, free, unequal, persistent, the smaller rounded or ovate, the longer lanceolate and bearing a short awn; margins minutely ciliate.
   COROLLA: Petals 3–4-lobed, tubular at the base, white or pinkish, ciliate on the lobes.
   STAMENS: 4, epipetalous; anthers white.
   PISTIL: Ovary inferior, ciliate; style 1, unbranched; stigma capitate.
FRUIT: A few-seeded capsule subtended by a persistent calyx.
COMMENTS: Frequent on sandy soils in the Coastal Prairies and Rio Grande Plains. The leaves are eaten by white-tailed deer, and the seeds are occasionally consumed by bobwhite quail and Rio Grande turkeys.

RUBIACEAE

Richardia L.
1a. Sepals 5–7; petals 5–7; leaves ovate to broadly rounded. *Richardia brasiliensis*
1b. Sepals 4; petals 4; leaves lanceolate. *Richardia tricocca*

TROPICAL MEXICAN CLOVER
Richardia brasiliensis Gomes

PERENNIAL
STEMS: Prostrate, mat-forming, not rooting at the nodes, purple-tinged, pubescent with hairs of 2 different lengths.
LEAVES: Simple, opposite, scabrous; blades ovate to broadly-rounded; margins entire, ciliate; petioles present; stipular bristles short.
INFLORESCENCE: Flowers 1-several in the leaf axils.
   CALYX: Sepals 5–7, slightly united at the base, lobes lanceolate, slightly unequal, pubescent, margins ciliate.
   COROLLA: Petals 5—7, united into a tube, white, ciliate from within the tube, actinomorphic.
   STAMENS: Usually 6, epipetalous, inserted in the cleft between the lobes; anthers white.
   PISTIL: Ovary inferior, ciliate; style 1, branches 3.
FRUIT: A capsule with a persistent calyx.
COMMENTS: Frequent on deep sandy soils in the Coastal Prairies and eastern portion of the Rio Grande Plains. The seeds are eaten by bobwhite quail and mourning doves. The leaves are eaten by Rio Grande turkeys.

RUBIACEAE

PRAIRIE BUTTONWEED
Richardia tricocca (T.&G.) Standl.
[Syn. *Diodia tricocca* T.&G.]

PERENNIAL
STEMS: Prostrate, mat-forming, pubescent.
LEAVES: Simple, opposite, sessile, mostly glabrous; blades lanceolate; margins entire with regularly spaced hairs; stipular bristles 3–4 mm long.
INFLORESCENCE: Flowers in clusters in the upper leaf axils.
    CALYX: Sepals 4, united near the base, lanceolate with long, stiff bristles on the margins.
    COROLLA: Petals 4, united more than 1/2 their length, actinomorphic, pubescent at the base of the throat.
    STAMENS: 4, epipetalous, attached at the base of the corolla lobes, white.
    PISTIL: Ovary inferior, bristly.
FRUIT: A bristly capsule often brownish-red, subtended by a persistent calyx; seeds 3–4.
COMMENTS: Frequent on loamy soils in prairies and openings in the Rio Grande Plains and Coastal Prairies. The leaves are occasionally eaten by bobwhite quail and white-tailed deer.

SCROPHULARIACEAE

PRAIRIE AGALINUS
Agalinus heterophylla (Nutt.) Small ex Britt.
[Syn. *Gerardia heterophylla* Nutt.]

ANNUAL: From a taproot.
STEMS: Erect, branching, slightly ribbed, glabrous, turning dark brown upon drying.
LEAVES: Simple, alternate, sessile, glabrous; blades linear with mealy scales; margins entire; apex acute or mucronate.
INFLORESCENCE: Flowers solitary from the upper leaf axils.
    CALYX: Sepals 5, united.
    COROLLA: Petals 5, united, zygomorphic, pink, pubescent on the outside.
    STAMENS: 4, epipetalous.
    PISTIL: Ovary superior; style 1, unbranched.
FRUIT: A globose capsule partially enclosed by the calyx.
COMMENTS: Frequent on various soils (usually moist) in fallow fields, prairies, plains, and grasslands in the Rio Grande Plains and Coastal Prairies. The leaves are eaten by white-tailed deer and cattle.

SOLANACEAE

1a. Calyx enlarged and inflated in fruit and completely enclosing the berry.
   2
1b. Calyx in fruit not greatly enlarged nor inflated. **3**

2a. Corolla yellow with 5 brownish-purple nectar guides near the base;
   calyx in fruit round. *Physalis*
2b. Corolla lavender-violet with green areas on the back; calyx in fruit
   5-angled. *Quincula*

3a. Flowers usually solitary at the nodes. *Chamaesaracha*
3b. Flowers in cymes or in racemose clusters. *Solanum*

Chamaesaracha Gray
1a. Stems pubescent with a mixture of long, straight hairs and smaller,
   stellate hairs. *Chamaesaracha coronopus*
1b. Stems villous but usually lacking smaller, stellate hairs. *Chamaesaracha
   sordida*

FALSE NIGHTSHADE
Chamaesaracha coronopus (Dun.) Gray

PERENNIAL
STEMS: Prostrate, pubescent with a mixture of long, straight hairs and
much smaller, stellate hairs.
LEAVES: Simple, alternate, nearly opposite above, pubescent; blades
oblanceolate to nearly ovate; margins lobed and extended down the
petioles.
INFLORESCENCE: Flowers usually solitary from the leaf axils; pedicels
pubescent, elongated.
   CALYX: Sepals 5, united about 3/4 their length, lobes densely pubescent.
   COROLLA: Petals 5, united, greenish-yellow, rotate with a dense cluster
   of hairs at the base, opening on sunny afternoons.
   STAMENS: 5, epipetalous.
   PISTIL: Ovary superior, orange, enclosed in a dense cluster of hairs;
   style 1, much longer than the stamens.
FRUIT: A berry with the calyx closely fitting around the fruit.
COMMENTS: Occasional on sand, clay and rocky soils in the Rio Grande
Plains and western Coastal Prairies. The leaves and stems are occasionally
eaten by white-tailed deer.

SOLANACEAE

HAIRY FALSE NIGHTSHADE
Chamaesaracha sordida (Dun.) Gray

PERENNIAL: Somewhat woody near the base.
STEMS: Densely villous.
LEAVES: Simple, alternate, villous, shorter hairs glandular, lanceolate, margins sinuate to lobed.
INFLORESCENCE: Flowers axillary with an elongated, villous pedicel.
   CALYX: Sepals, 5, united, actinomorphic, purple-tinged, with 5 separate tufts of white hairs near the base of stamen filaments.
   STAMENS: Stamens 5, epipetalous, anthers yellow.
   PISTIL: Ovary superior, style 1, unbranched.
FRUIT: A small berry.
COMMENTS: Occasional on sand, clay, or rocky soils in the Rio Grande Plains. The leaves and stems are eaten by white-tailed deer.

GROUND CHERRY
Physalis cinerascens (Dun.) A.S. Hitchc.
[Syn. *Physalis viscosa* L. var. *cinerascens* (Dun.) Waterfall] [Syn. *Physalis viscosa* L. var. *spathulifolia* (Torr.) Gray]

ANNUAL or PERENNIAL
STEMS: Erect or sprawling, with minute, stellate pubescence.
LEAVES: Simple, alternate, stellate-pubescent; blades ovate or spatulate; margins entire or wavy; petioles stellate-pubescent.
INFLORESCENCE: Flowers solitary, turning downward, arising from the leaf axils, pedicels stellate-pubescent.
   CALYX: Sepals 5, united about 3/4 their length; in flower about one-half the length of the corolla; in fruit inflated, and enclosing the berry; stellate-pubescent.
   COROLLA: Petals 5, united, rotate, yellow, with 5, brownish-purple, pubescent nectar guides near the base.
   STAMENS: 5, epipetalous; anthers lavender.
   PISTIL: Ovary superior; style 1, unbranched.
FRUIT: A berry.
COMMENTS: Frequent on deep sands in prairies and openings in the Rio Grande Plains and Coastal Prairies. The leaves and fruits are important foods of Rio Grande turkeys, white-tailed deer, and javelinas. The seeds are eaten by bobwhite quail and mourning doves. The leaves are also occasionally eaten by cattle.

SOLANACEAE

PURPLE GROUNDCHERRY
Quincula lobata (Torr.) Raf.
[Syn. *Physalis lobata* Torr.]

PERENNIAL
STEMS: Sprawling, branching from the base, angled, glabrous.
LEAVES: Simple, alternate, with scattered, white vescicles; blades ovate to ovate-lanceolate; margins wavy and extended to the petiole.
INFLORESCENCE: Flowers 1–2 from the leaf axils.
    CALYX: Sepals 5, united for about two-thirds of their length, covered with white vescicles.
    COROLLA: Petals 5, united, rotate, lavender-violet with a lance-shaped green configuration on the back, pubescent at base.
    STAMENS: 5, epipetalous.
    PISTIL: Ovary superior; style 1, unbranched; stigma green.
FRUIT: A berry enclosed within a 5-angled, inflated calyx.
COMMENTS: Infrequent on clays and loams in chaparral and brushlands in the western half of the Rio Grande Plains. The leaves are occasionally eaten by white-tailed deer.

SOLANACEAE

Solanum L.
1a. Plants spiny throughout. **2**
1b. Plants glabrous or pubescent but not spiny. **3**

2a. Corolla yellow; leaves pinnatisect; fruit a spiny capsule. *Solanum rostratum*
2b. Corolla purple or occasionally white; leaf margins usually entire; fruit a yellow berry turning black. *Solanum elaegnifolium*

3a. Plants annuals; fruit a purple or black berry; stems pubescent. *Solanum ptycanthum*
3b. Plants perennials; fruit a red berry; stems glabrous. *Solanum triquetrum*

SILVERLEAF NIGHTSHADE, TROMPILLO
Solanum elaeagnifolium Cav.

PERENNIAL: Often slightly woody near the base.
STEMS: Erect, usually with scattered prickles, canescent, grayish, with scurfy pubescence on new growth.
LEAVES: Simple, alternate, grayish-green, stellate-pubescent; blades lanceolate; margins entire; apex obtuse; petioles present.
INFLORESCENCE: A few-flowered cyme; peduncles and pedicels stellate-pubescent and often with scattered, brown prickles.
   CALYX: Sepals 5, united for about half their length, about one-third the length of the corolla, stellate-pubescent with scattered prickles.
   COROLLA: Petals 5, united, spreading laterally, purple or occasionally white; with yellow nectar guides.
   STAMENS: 5, epipetalous, attached at the base of the corolla; anthers poricidal, yellow.
   PISTIL: Ovary superior, stellate-pubescent; style 1, unbranched, much longer than the stamens.
FRUIT: A yellow, marble-sized berry, turning black.
COMMENTS: Common on various soils in prairies, openings, waste places, and fields in the Rio Grande Plains and Coastal Prairies.
The fruits are eaten by white-tailed deer, javelinas and feral pigs, and the seeds are eaten by bobwhite quail.

SOLANACEAE

BLUEFLOWER BUFFALO-BUR, AMERICAN NIGHTSHADE, HIERBA MORA NEGRA
Solanum ptycanthum Dun. ex DC.
[Syn. *Solanum americanum* of Texas authors, not Mill.] [Syn. *Solanum nigrum* of Texas authors, not L.]

ANNUAL: From a taproot.
STEMS: Erect, branching, slightly angled, pubescent with antrorsely bent hairs.
LEAVES: Simple, mostly alternate, pubescent with more hairs on the lower epidermis; blades ovate; margins entire to slightly wavy and extending as minute wings along the petioles; apex obtuse or acute; petioles pubescent.
INFLORESCENCE: A cyme from the leaf axils; peduncle and pedicels pubescent.
    CALYX: Sepals 5, united, slightly unequal, pubescent.
    COROLLA: Petals 5, united, lanceolate, white or blue-lavender, margins minutely pubescent.
    STAMENS: 5, epipetalous; anthers yellow, poricidal.
    PISTIL: Ovary superior; style 1, unbranched, pubescent.
FRUIT: A purple to black berry; calyx persistent; lobes slightly recurved.
COMMENTS: Frequent on sandy or rocky soils in open woods, thickets, and openings in the Rio Grande Plains. The leaves are eaten by chachalacas, while the fruits are eaten by several species of passerine birds.

BUFFALOBUR, KANSAS THISTLE, MALA MUJER
Solanum rostratum Dun.

ANNUAL: From a taproot.
STEMS: Sprawling; all parts of plants with formidable, long, yellow, stout spines, some with purple bases; pubescent with stellate hairs.
LEAVES: Simple, alternate, spiny, stellate-pubescent; blades pinnatisect; spines on larger veins.
INFLORESCENCE: A raceme; stellate-pubescent pedicles with stout spines.
    CALYX: Sepals 5, united, with narrowly lanceolate lobes, spiny and pubescent.
    COROLLA: Petals 5, united, actinomorphic, golden-yellow.
    STAMENS: 5, epipetalous; filaments short; anthers poricidal.
    PISTIL: Ovary superior; style 1, unbranched.
FRUIT: A spiny, pubescent capsule.
COMMENTS: Occasional on various soils in waste places and along roads in the Rio Grande Plains and Coastal Prairies. The seeds are eaten by bobwhite quail and Rio Grande turkeys.

SOLANACEAE

TEXAS NIGHTSHADE, FALSE WILD PEPPER, HIERBA MORA
Solanum triquetrum Cav.

PERENNIAL
STEMS: Erect or usually scandent, glabrous.
LEAVES: Simple, alternate, nearly glabrous; blades lanceolate to ovate; margins often with a few scattered hairs; apex acute or obtuse.
INFLORESCENCE: A few-flowered cyme.
    CALYX: Sepals 5, united.
    COROLLA: Petals 5, united, white.
    STAMENS: 5, epipetalous; anthers poricidal, bright yellow.
    PISTIL: Ovary superior; style 1, unbranched.
FRUIT: A red berry.
COMMENTS: Common on various soils in thickets, woods, pastures, and waste places in the Rio Grande Plains and Coastal Prairies. The leaves are frequently eaten by javelinas, and the fruits are eaten by several species of passerine birds.

STERCULIACEAE

FLORIDA WALTHERIA, HIERBA DEL SOLDADO
Waltheria indica L.
[Syn. *Waltheria americana* L.]

PERENNIAL: From a woody rootstock.
STEMS: Erect, usually not more than 0.5 m tall; densely stellate-pubescent.
LEAVES: Simple, alternate, pubescent; blades ovate or elliptic, venation pinnate, veins conspicuously elevated below; margins dentate; apex obtuse; petioles present; stipules linear.
INFLORESCENCE: In dense, headlike clusters in the leaf axils; linear bracts present in the axis.
    CALYX: Sepals 5, united for about two-thirds their length, lobes acuminate, pubescent, margins with glandular hairs.
    COROLLA: Petals 5, free, slightly exceeding the length of the calyx, yellow; apex rounded.
    STAMENS: 5; monadelphous.
    PISTIL: Ovary superior; style branches numerous.
FRUIT: A pubescent capsule.
COMMENTS: On sandy or rocky soil in the extreme southern portion of the Rio Grande Plains. The leaves are occasionally eaten by cattle.

TYPHACEAE

NARROW-LEAF CATTAIL, TULE
Typha domingensis Pers.

PERENNIAL: From an extensive rhizome system; forming large colonies; monoecious.

STEMS: Erect, glabrous, to 2 m tall.

LEAVES: Simple, alternate, venation parallel, glabrous; blades strap-shaped, sheathing the stem, flat to slightly rounded below, lacking a distinct midvein.

INFLORESCENCE: An interrupted spike with the staminate flowers above the pistillate. A clear demarcation zone occurs between the male and female flowers.

STAMINATE FLOWERS: Perianth reduced to minute bracts; inflorescence covered with a dense mass of yellow pollen; flowers falling soon after anthesis and leaving a persistent, naked rachis.

PISTILLATE FLOWERS: Perianth minute, brownish, persistent on the rachis.

FRUIT: Minute, plumed nutlets.

COMMENTS: Common in brackish or fresh marshes and pools in the Rio Grande Plains and Coastal Prairies. The leaves are occasionally eaten by white-tailed deer; the rhizomes are eaten by feral pigs.

URTICACEAE

1a. Leaves opposite; plants with stinging hairs. *Urtica*
1b. Leaves alternate; plants usually pubescent, but stinging hairs absent.
   Parietaria

PENNSYLVANIA PELLITORY
Parietaria pennsylvanica Muhl. ex Willd.

ANNUAL: From a shallow taproot; usually in colonies; polygamous.
STEMS: Low-growing, pubescent, with a watery juice.
LEAVES: Simple, alternate, pubescent; blades lanceolate; margins entire.
INFLORESCENCE: Flowers in axillary clusters with small involucral bracts subtending the calyx.
STAMINATE FLOWERS:
   CALYX: Sepals 4, united, pubescent.
   COROLLA: Petals absent.
   STAMENS: 4.
PISTILLATE FLOWERS:
   CALYX: Similar to the staminate flowers.
   STAMENS: Absent.
   PISTIL: Ovary superior; style 1.
FRUIT: A brown, lustrous achene.
COMMENTS: Frequent on various soils in pastures and woods in the Rio Grande Plains and Coastal Prairies. The leaves are occasionally eaten by white-tailed deer.

HEART-LEAF NETTLE, ORTIGUILLA
Urtica chamaedryoides Pursh

ANNUAL: From a shallow taproot; monoecious or dioecious.
STEMS: Erect; stems and leaves with white-translucent, stinging hairs; spines swollen at the base and tapered into a sharp, capillary apex.
LEAVES: Simple, opposite, spiny; blades ovate to ovate-lanceolate; margins serrate; apex obtuse; petioles present.
INFLORESCENCE: Flowers in globular heads.
   CALYX: Sepals 4, free, with stinging hairs.
   COROLLA: Petals absent.
STAMINATE FLOWERS: Stamens 4; pollen white.
PISTILLATE FLOWERS: Ovary superior.
FRUIT: An achene.
COMMENTS: Common on various soils in pastures, woods and waste places in the Rio Grande Plains and Coastal Prairies. The young leaves are eaten by feral pigs.

VERBENACEAE

1a. Plants prostrate, trailing and mat-forming; inflorescence a head; corolla white or rose-colored with a yellow center. *Phyla*
1b. Plants erect; inflorescence a spike; corolla blue or lavender. *Verbena*

Phyla Lour.
1a. Leaves usually linear to linear-lanceolate, marginal teeth usually present on the upper three-quarters of the blade. *Phyla incisa*
1b. Leaves usually rhombic, cuneate, or spatulate; marginal teeth usually present throughout. *Phyla nodiflora*

SAWTOOTH FROG-FRUIT
Phyla incisa Small
[Syn. *Lippia incisa* (Small) Tidest.]

PERENNIAL: Often slightly woody at the base.
STEMS: Prostrate, trailing, mat-forming; minutely appressed-pubescent.
LEAVES: Simple, opposite, pubescent; blades linear to linear-lanceolate; margins with teeth usually on the upper three-quarters of the blade; petioles present.
INFLORESCENCE: A cylindrical spike from a long peduncle arising from the leaf axils.
    CALYX: Sepals of 2–4 segments, inconspicuous, membranous, pubescent.
    COROLLA: Petals 5, united, zygomorphic, white, rose-colored with a yellow center; tubular below, about 4.0–4.5 mm long.
    STAMENS: 4, epipetalous.
    PISTIL: Ovary superior; style 1, unbranched; stigma capitate.
FRUIT: Small, dry, nut-like.
COMMENTS: *Phyla incisa* is frequent on various soils in prairies, openings, waste places, and along roads in the Rio Grande Plains and Coastal Prairies. The leaves are eaten by white-tailed deer, javelinas, feral pigs and cattle.

TURKEY TANGLE, COMMON FROG-FRUIT
Phyla nodiflora (L.) Greene
[Syn. *Lippia nodiflora* (L.) Michx.]

Similar to *P. incisa*. However, it usually has broader leaves and the teeth are widely spaced over most of the marginal surface. It is frequent on wet or moist soil in a variety of habitats in the Coastal Prairies and Rio Grande Plains. The leaves are eaten by white-tailed deer and javelinas.

VERBENACEAE

Verbena L.
1a. Leaf blades pinnatisect below, linear above. *Verbena officinale*
1b. Leaf blades ovate, margins lobed, not pinnatisect. *Verbena plicata*

SLENDER VERVAIN, TEXAS VERVAIN
Verbena officinale L. ssp. *halei* (Small) Barber
[Syn. *Verbena halei* Small]

PERENNIAL: From a woody taproot.
STEMS: Erect, 4-angled, with widely scattered hairs on the ribs.
LEAVES: Simple, opposite, strigose-pubescent; blades pinnatisect below, linear above, midvein conspicuous; margins revolute.
INFLORESCENCE: A spike; each flower subtended by a strigose, lanceolate bract.
    CALYX: Sepals 5, united, strigose, about half the length of the corolla, apex acute.
    COROLLA: Petals 5, united, blue or lavender, pubescent within the tube at the juncture of the upper lobes and the throat below, about 5 mm long.
    STAMENS: 4, epipetalous.
    PISTIL: Ovary superior.
FRUIT: 4, cylindrical, nut-like structures.
COMMENTS: Common on various soils in prairies, openings, fields, and waste places in the Rio Grande Plains and Coastal Prairies. The leaves and stems are eaten by white-tailed deer and cattle. The leaves and seeds are utilized by Rio Grande turkeys.

FANLEAF VERVAIN
Verbena plicata Greene

PERENNIAL: From a taproot.
STEMS: Erect, densely pubescent with numerous glandular and aglandular hairs.
LEAVES: Simple, opposite, densely pubescent, conspicuously net-veined; blades ovate; margins lobed and extending along the petiole to the axis.
INFLORESCENCE: A terminal spike with a lanceolate bract subtending each flower; bracts longer than calyx.
    CALYX: Sepals 5, united, glandular-pubescent, about 4 mm long.
    COROLLA: Petals 5, slightly zygomorphic, blue, pubescent on the outside.
    STAMENS: 4, epipetalous, included within the corolla tube.
    PISTIL: Ovary superior; style 1.
FRUIT: 4-lobed, nut-like structures.
COMMENTS: Frequent on caliche or dry loams in the Rio Grande Plains. The leaves are occasionally eaten by white-tailed deer.

VIOLACEAE

NOD-VIOLET, GREEN VIOLET
Hybanthus verticillatus (Ort.) Baill.

PERENNIAL: Forming small colonies.
STEMS: Low-growing, minutely-scabrous.
LEAVES: Simple, alternate, pubescent; blades linear to narrowly lanceolate; margins entire; stipules present.
INFLORESCENCE: Flowers solitary from the leaf axils; pedicels reflexed; flowers turned downward.
    CALYX: Sepals 5, free, pubescent.
    COROLLA: Petals 5, united near the base, zygomorphic; 4 smaller petals greenish-white, lavender near the base, margins erose; larger petal green below, white above, flared near the apex, twice as long as the other petals.
    STAMENS: 5, united in a sheathlike structure around the pistil; sheath brown, membranous above.
    PISTIL: Ovary superior; style 1, unbranched.
FRUIT: A capsule.
COMMENTS: Frequent on various soils in openings and waste places in the Coastal Prairies and Rio Grande Plains. The leaves are eaten by white-tailed deer, cattle, and bobwhite quail.

VITACEAE

1a. Leaves simple, succulent; plants foul-smelling; petals 4. *Cissus*
1b. Leaves bi- or tripinnately compound, not succulent nor with a foul odor; petals 5. *Ampelopsis*

PEPPERVINE
Ampelopsis arborea (L.) Koehne

PERENNIAL
STEMS: Climbing or sprawling, slightly ribbed, glabrous, often purple-tinged.
LEAVES: Bi- or tripinnately compound; alternate, but often with a twining tendril opposite a leaf; pubescent; leaflets toothed on the margins; apex acuminate; petioles and rachis pubescent.
INFLORESCENCE: Flowers in paniculate or cymose clusters arising from the upper nodes; branches pubescent.
    CALYX: Sepals 5, united into a shallow cup, pubescent.
    COROLLA: Petals 5, united at the base, actinomorphic; lobes spreading laterally.
    STAMENS: 5, free, attached below a cuplike disk; anthers yellow.
    PISTIL: Ovary superior; style 1, unbranched.
FRUIT: A black berry similar to a grape.
COMMENTS: Occasional along streams and in bottom woods in the Coastal Prairies and eastern and extreme southern portions of the Rio Grande Plains. The seeds are occasionally eaten by bobwhite quail.

POSSUM-GRAPE, IVY TREEBINE, HIERBA DEL BUEY
Cissus incisa (T.&G.) Des Moulins

PERENNIAL: From a woody base; foul-smelling.
STEMS: A climbing vine with tendrils, glabrous.
LEAVES: Simple, alternate, glabrous, succulent; blades usually trilobed; margins broadly toothed; petioles and stipules present.
INFLORESCENCE: Flowers in umbellate or corymbose cymes.
    CALYX: Sepals 4, united, minute.
    COROLLA: Petals 4, free, yellowish-green, actinomorphic.
    STAMENS: 4, free; anthers yellow.
    PISTIL: Ovary superior; style 1, unbranched.
FRUIT: A black berry.
COMMENTS: Common on various soils in pastures and woods in the Rio Grande Plains and Coastal Prairies. The leaves are occasionally eaten by white-tailed deer, and both the leaves and seeds are consumed by bobwhite quail.

ZYGOPHYLLACEAE

CALTROP
Kallstroemia californica (S. Wats.) Vail

ANNUAL: From a taproot.
STEMS: Prostrate, pubescent with 2 types of hairs, the shorter more numerous, bent; the longer hairs straight and perpendicular to the stem.
LEAVES: Even-pinnately compound, opposite, pubescent; leaflets usually 4 pairs per leaf, 1.5–2.0 cm long; apex rounded; stipules pubescent.
INFLORESCENCE: Flowers usually solitary from the leaf axils; pedicels pubescent.
CALYX: Sepals 5, free, about two-thirds the length of the corolla, pubescent.
COROLLA: Petals 5, free, yellow-orange.
STAMENS: 10, free; anthers yellow.
PISTIL: Ovary superior; style 1, unbranched.
FRUIT: An appressed-pubescent capsule with a persistent style, tuberculed but not spiny.
COMMENTS: Frequent in prairies, openings, and waste places in the Rio Grande Plains and Coastal Prairies. The seeds are occasionally eaten by bobwhite quail.

BIBLIOGRAPHY

Ajilvsgi, G. 1984. Wildflowers of Texas. Shearer Publishings, Bryan, TX. 414 pp.

Arnold, L. A., Jr. 1976. Seasonal food habits of white-talled deer (*Odocoilcus virginianus* Bod.) on the Zachry Ranch in south Texas. Unpublished M.S. thesis, Texas A&I Univ., Kingsville. 59 pp.

Arnold, L. A., Jr. and D. L. Drawe. 1979. Seasonal food habits of white-tailed deer in the South Texas Plains. J. Range Manage., 32:175-178.

Baskett, T. S., M. W. Sayre, R. E. Tomlinson, and R. W. Mirarchi. 1993. Ecology and management of the mourning dove. Stackpole Books, Harrisburg, PA. 567 pp.

Beasom, S. L. 1973. Ecological factors affecting wild turkey reproductive success in south Texas. Unpublished Ph.D. dissertation, Texas A&M Univ., College Station. 215 pp.

Chamrad, A. D. 1966. Winter and spring food habits of white-tailed deer on the Welder Wildlife Refuge. M.S. thesis, Texas Technological Coll., Lubbock. 181 pp.

Clover, E. U. 1937. Vegetational survey of the lower Rio Grande Valley, Texas. Madroño, 4:41-66; 77-100.

Conant, R. 1975. A field guide to reptiles and amphibians of eastern and central North America. Houghton Mifflin Company, Boston. 429 pp.

Correll, D. S. and M. C. Johnston. 1970. Manual of the vascular plants of Texas. Texas Research Foundation, Renner, TX. 1881 pp.

Davis, R. B. 1951. The food habits of white-tailed deer on the cattle-stocked, live oak-mesquite ranges of the King Ranch, as determined by analyses of deer rumen contents. Unpublished M.S. thesis, Texas A&M Coll., College Station. 97 pp.

Davis, R. B. 1952. A study of some interrelationships of a native south Texas range, its cattle, and its deer. Unpublished Ph.D. dissertation, Texas A&M Coll., College Station. 114 pp.

Davis, R. B. and C. K. Winkler. 1968. Brush vs. cleared range as deer habitat in southern Texas. J. Wildl. Manage., 32:321-329.

Davis, W. B. 1974. The mammals of Texas. Texas Parks and Wildlife Department, Bull. 41, rev. ed., Austin. 267 pp.

Drawe, D. L. 1967. Seasonal forage preferences of deer and cattle on the Welder Wildlife Refuge. M.S. thesis, Texas Technological Coll., Lubbock. 97 pp.

Drawe, D. L. 1968. Mid-summer diet of deer on the Welder Wildlife Refuge. J. Range Manage., 21:164-166.

Drawe, D. L. and T. W. Box. 1968. Forage ratings for deer and cattle on the Welder Wildlife Refuge. J. Range Manage., 21:225-228.

Everitt, J. H. 1972. Spring food habits of white-tailed deer (*Odocoileus virgianus* Bod.) on the Zachry Ranch in South Texas. M.S. thesis, Texas A&I Univ., Kingsville. 114 pp.

Everitt, J. H. 1986. Nutritive value of fruits or seeds of 14 shrub and herb species from South Texas. Southwestern Nat., 31:101-104.

Everitt, J. H. and M. A. Alaniz. 1980. Fall and winter diets of feral pigs in South Texas. J. Range Manage., 33:126-129.

Everitt, J. H. and D. L. Drawe. 1974. Spring food habits of white-tailed deer in the South Texas Plains. J. Range Manage., 27:15-20.

Everitt, J. H. and D. L. Drawe. 1993. Trees, shrubs, and cacti of South Texas. Texas Tech Univ. Press, Lubbock. 213 pp.

BIBLIOGRAPHY

Everitt, J. H. and C. L. Gonzalez. 1979. Botanical composition and nutrient content of fall and early winter diets of white-tailed deer in south Texas. Southwestern Nat., 24:297-310.

Everitt, J. H. and C. L. Gonzalez. 1981. Seasonal nutrient content in food plants of white-tailed deer on the South Texas Plains. J. Range Manage., 34:506-510.

Everitt, J. H., C. L. Gonzalez, M. A. Alaniz, and G. V. Latigo. 1981. Food habits of the collared peccary on South Texas rangelands. J. Range Manage., 34:141-144.

Everitt, J. H., C. L. Gonzalez, G. Scott, and B. E. Dahl. 1981. Seasonal food preferences of cattle on native range in the South Texas Plains. J. Range Manage., 34:384-388.

Fulbright, T. E. and A. Garza, Jr. 1991. Forage yield and white-tailed deer diets following live oak control. J. Range Manage. 44:451-455.

Fulbright, T. E., J. P. Reynolds, and S. L. Beasom. 1993. Effects of browse rejuvenation on white-tailed deer diets and nutrition. Texas J. Agric. Nat. Resources. 6:41-48.

Gould, F. W. 1975. Texas plants—A checklist and ecological summary. MP-585, Texas Agricultural Experiment Station, Texas A&M Univ., College Station. 121 pp.

Guthery, F. S. 1975. Food habits of sandhill cranes in southern Texas. J. Wildl. Manage. 39:221-223.

Hatch, S. L., K. N. Gandhi, and L. E. Brown. 1990. Checklist of the vascular plants of Texas. MP-1655. Texas Agricultural Experiment Station, Texas A&M Univ., College Station. 158 pp.

Hatch, S. L. and J. Pluhar. 1993. Texas range plants. Texas A&M Univ. Press, College Station. 326 pp.

Higginbotham, I., Jr. 1975. Composition and production of vegetation on the Zachry Ranch in the South Texas Plains. M.S. thesis, Texas A&I Univ., Kingsville. 131 pp.

Hoffman, D. M. 1962. The wild turkey in eastern Colorado. Dept. Project W-96-D. Colorado Game and Fish Department. 47 pp.

Johnston, M. C. 1955. Vegetation of the eolian plain and associated coastal features of southern Texas. Unpublished Ph.D. dissertation, Univ. of Texas, Austin. 167 pp.

Johnston, M. C. 1963. Past and present grasslands of southern Texas and northeastern Mexico. Ecology, 44:456-466.

Jones, F. B. 1975. Flora of the Texas coastal bend. Welder Wildlife Foundation Contribution B-6, Rob and Bessie Welder Wildlife Foundation, Sinton, Texas. 262 pp.

Jones, F. B., C. M. Rowell, Jr., and M. C. Johnston. 1961. Flowering plants and ferns of the Texas coastal bend counties. Welder Series B-1, Rob and Bessie Welder Wildlife Foundation, Sinton, Texas. 146 pp.

Jones, J. K., Jr., R. S. Hoffman, D. W. Rice, C. Jones, R. J. Baker, and M. D. Engstrom. 1992. Revised checklist of North American mammals north of Mexico, 1991. Occas. Papers Mus., Texas Tech Univ., 146:1-23.

Latham, R. M. 1956. Complete book of the wild turkey. The Stockpole Co., Harrisburg, Pennsylvania. 265 pp.

Lehmann, V. W. 1984. Bobwhites in the Rio Grande Plain of Texas. Texas A&M Univ. Press, College Station. 371 pp.

BIBLIOGRAPHY

Litton, G. W. 1977. Food habits of the Rio Grande turkey in the Permian Basin of Texas. Technical Series Bull. No. 18. Texas Parks and Wildlife Department, Austin. 22 pp.

Lonard, R. I. 1993. Guide to grasses of the Lower Rio Grande Valley. Univ. of Texas-Pan American Press, Edinburg. 240 pp. (with illustrations by N. A. Browne and A. L. Egle.)

Lonard, R. I., J. H. Everitt, and F. W. Judd. 1991. Woody plants of the Lower Rio Grande Valley, Texas. Texas Memorial Museum, Univ. of Texas Press, Austin. 179 pp.

Lonard, R. I. and F. W. Judd. 1980. Phytogeography of South Padre Island, Texas. Southwestern Nat., 25:313-322.

Lonard, R. I. and F. W. Judd. 1981. The terrestrial flora of South Padre Island, Texas. Texas Memorial Museum, Misc. Papers, 6:1-74.

Lonard, R. I., F. W. Judd, and S. L. Sides. 1978. Annotated checklist of the flowering plants of South Padre Island, Texas. Southwestern Nat., 23:497-510.

Lukefahr, M. J. and D. F. Martin. 1962. A native host plant of the boll weevil and other cotton insects. J. Econ. Entomol., 55:150-155.

Marion, W. R. 1976. Plain Chachalaca food habits in south Texas. Auk, 93:376-379.

Pattee, O. H. 1977. Effects of nutrition on wild turkey reproduction in south Texas. Ph.D. dissertation. Texas A&M Univ., College Station. 126 pp.

Rappole, J. H. and G. W. Blacklock. 1985. Birds of the Texas Coastal Bend; abundance and distribution. Texas A&M Univ. Press, College Station. 126 pp.

Richardson, A. 1990. Plants of southernmost Texas. Gorgas Science Foundation, Brownsville, TX. 298 pp.

Richardson, A. 1995. Plants of the Rio Grande Delta. Univ. of Texas Press, Austin. 332 pp.

Runyon, R. 1947. Vernacular names of plants indigenous to the lower Rio Grande Valley of Texas. The Brownsville News Publishing Co., Brownsville, Texas. 24 pp.

Scifres, C. J., J. L. Mutz, and G. P. Durham. 1976. Range improvement following chaining of south Texas mixed brush. J. Range Manage., 29:418-421.

Sperry, O. E., J. W. Dollahite, G. O. Hoffman, and B. J. Camp. 1968. Texas plants poisonous to livestock. B-1028, Texas Agricultural Experiment Station, Texas A&M Univ., College Station. 57 pp.

Vines, R. A. 1960. Trees, shrubs, and woody vines of the southwest. Univ. of Texas Press, Austin. 1104 pp.

Walmo, O. C. 1956. Ecology of scaled quail in west Texas. Special Project W-57-R. Texas Game and Fish Commision, Austin. 134 pp.

Wilson, M. H. 1984. Comparative ecology of bobwhite and scaled quail in southern Texas. Ph.D. dissertation, Oregon State University, Corvallis. 85 pp.

Wilson, M. H. and J. A. Crawford. 1987. Habitat selection by Texas bobwhites and chestnut-bellied scaled quail in south Texas. J. Wildl. Manage., 51:575-582.

GLOSSARY

acaulescent: Stemless or stems subterranean.

achene: A small, dry, indehiscent, one-seeded fruit with a thin ovary wall free from the seed.

actinomorphic: A flower with radial symmetry (regular).

acuminate: Long-pointed; tapering to an elongated point.

adnate: Fusion of unlike parts; usually the attachment of the stamen to the corolla.

alternate: Leaves borne singly at each node.

annual: A plant that germinates, flowers, produces seeds and dies in the same year.

anther: The expanded, pollen-bearing portion of the stamen.

anthocarp: A fusion of the fruit with the perianth or receptacle.

antrorse: Directed upward or forward.

apetalous: Lacking petals.

apex: The point farthest from the point of attachment; the tip.

appendage: An attached structure.

appressed: Pressed closely against a structure.

arenchyma: With well-developed air spaces throughout.

auricle: A small, ear-shaped appendage.

auriculate: With auricles.

awl: Narrow, stiff, and sharp-pointed.

awn: Stiff, needle-like extension.

axil: The upper angle between the stem and the leaf.

berry: A fleshy fruit with more than one seed. The seeds are embedded in pulpy tissue; e. g., a tomato or a grape.

biennial: Completing the life cycle in two growing seasons.

bifid: Deeply two-lobed.

bilabiate: Two-lipped; zygomorphic.

bipinnately compound: Twice-pinnately compound.

blade: The expanded, laminar portion of a leaf.

bract: A reduced or modified leaf often associated with a flower or an inflorescence.

bulb: An underground stem with thickened, fleshy scales; e. g., an onion.

calyx: Collectively, the sepals of a flower; the outermost series of floral parts in the perianth.

caliche: A calcium carbonate crust on stony soils in dry regions.

canescent: Gray or white due to a covering of short hairs.

capillary: A slender, hair-like structure.

capitate: Head-shaped.

capsule: A simple, dry fruit made up of more than one carpel and dehiscent at maturity; several to many-seeded.

carpel: A simple pistil, or one of the modified structures that forms a compound pistil.

caruncle: A protuberance near the seed scar.

chaff: A thin, dry scale that arises from the receptacle in some members of the Asteraceae.

cilia: See below.

ciliate: Hairs found on the margins of a structure.

circumscissile: Dehiscence so that the top separates like a lid.

clawed: The narrowed base of some sepals or petals.

cleft: Cut into lobes that extend more than halfway to the midrib.

compound: Composed of two or more similar elements; more than 1 blade per petiole.

connate: Fusion of like parts; e.g., stamens.

cordate: Heart-shaped.

corm: A short, solid, vertical, underground stem with papery leaves.

corolla: Petals of a flower.

corymb: A short, broad, more-or-less flat-topped, indeterminate inflorescence with pedicels of different lengths.

GLOSSARY

corymbose: Corymb-like.

cm: Centimeter; 0.01 m.

crenate: A leaf margin with blunt or rounded teeth.

crisped: Wavy or curled.

cyathium: An inflorescence consisting of a cup-like involucre containing a single pistil and male flowers with a single stamen; *Euphorbia*.

cyme: An inflorescence in which the central flower of a group is the most mature; a broad. usually flat-topped cluster of flowers.

deciduous: Falling off; not persistent.

dehiscent: Splitting open at maturity.

deltoid: Shaped like an equilateral triangle.

dentate: A leaf margin with sharp teeth pointing outward.

diadelphous: Stamens united into two unequal sets by their filaments; usually in a 9+1 arrangement.

dichotomous: Branched into two equal forks.

dioecious: Pistillate and staminate flowers on separate plants.

discoid head: A head with ray florets lacking.

disk florets: Tubular flowers of the Asteraceae.

eliasome: A sticky substance on the surface of a seed; a food source for ants.

elliptic: An ellipse; longer than wide and rounded at both ends.

entire: An even or smooth margin; lacking teeth.

epipetalous: Stamens attached to the corolla.

erose: Irregularly toothed margins.

even-pinnately compound: The terminal leaflets paired.

exserted: Protruding; usually structures extended beyond the length of the corolla.

fascicle: A cluster.

fibrous: Roots numerous and similar in diameter.

filament: The thread-like structure of a stamen that supports the anthers.

follicle: A dry fruit of one carpel splitting open along a single, ventral seam at maturity.

fruit: A ripened ovary and any associated structures that ripen with the ovary.

glabrous: Lacking hairs.

gland: A secreting organ; in plants, glands are often at the tips of hairs.

glaucous: Covered with a white substance that rubs off after polishing.

glomerules: Dense clusters.

glutinous: Covered with a sticky, glue-like substance.

head: Usually a dense inflorescence of sessile flowers on a short or broadened axis.

herbaceous: Not woody.

hirsute: Pubescent with coarse, stiff hairs.

hispid: Rough, with stiff, firm hairs.

hyaline: Thin, membranous and transparent or translucent.

hypanthium: A floral tube.

indehiscent: Not opening by splitting along regular lines, or not opening.

inferior ovary: An ovary positioned below the calyx.

inflorescence: A flower cluster.

imbricate: Overlapping like shingles on a roof.

involucel: A secondary set of sepal-like structures slightly below the flower.

involucre: A whorl of bracts subtending a flower or group of flowers.

keel: A conspicuous longitudinal ridge.

lanceolate: Lance-shaped; much longer than broad; edges curved along the broad portion.

latex: A milky juice.

leaflet: A single division of a compound leaf.

legume: A dry, dehiscent fruit of one carpel and two valves; seeds attached along a ventral suture; fruit of the Fabaceae.

GLOSSARY

ligulate: Strap-shaped; e. g., ray florets of the Asteraceae.

linear: Long and narrow with more-or-less parallel sides; resembles a line; e.g., a blade of grass.

lobe: A segment of an organ; a division to about the middle.

loment: A legume with constrictions between the seeds.

lyrate: Pinnatifid with the terminal lobe large and rounded and the lower lobes much reduced.

maculate: A blotch or spot.

malpighiaceous: T-shaped hairs.

margin: The edge of a leaf.

mericarp: A section of a schizocarp; one of the two halves of fruit of the Apiaceae.

mm: Millimeter=.001 m; 10mm=1cm.

monadelphous: Stamens that are united into one group by their filaments; common in the Malvaceae.

monoecious: Pistillate and staminate flowers on the same plant.

motte: A small grove of trees on a prairie.

mucronate: Tipped with a short spine.

node: A point on a stem which bears a leaf or leaves.

nutlet: A small nut; achene-like; a small, hard, indehiscent, one-seeded fruit.

obcordate: A leaf with a notched apex; inversely cordate.

oblanceolate: inversely lanceolate.

oblong: Elongated with roughly parallel sides; the length about two or three times the width.

obovate: Inversely ovate.

obtuse: Blunt or rounded at the apex.

ocrea: A sheath around the stem formed by the stipules.

odd-pinnately compound: With a terminal leaflet.

opposite: Leaves in pairs; on the opposite sides of the stem.

ovary: Ovule-bearing, basal portion of a pistil.

ovate: Egg-shaped; the broadest portion below the middle.

palmate venation: Divided from a common point; like the fingers of a hand.

palmately compound: With the leaflets palmately arranged.

panicle: An indeterminate-branching raceme; an inflorescence that is branched and rebranched.

paniculate: With flowers in panicles.

papilionaceous: An irregular corolla of some members of the Fabaceae; zygomorphic.

papillae: A short, rounded projection or bump.

papillose: With minute papillae.

pappus: Modified, scale-like, or bristly calyx of the Asteraceae.

pedicel: The stalk of a single flower.

peduncle: The stalk of a cluster of flowers.

peltate: Attached by the lower surface and not at the leaf margin.

perennial: A plant that continues to live for a number of years.

perfoliate: A leaf with the margins entirely surrounding the stem.

perianth: The calyx and corolla of a flower.

perigynous: With sepals, petals, and stamens attached to a calyx tube; surrounding but not attached to a superior ovary.

petal: One of the inner leaf-like parts of the flower; usually brightly colored.

petaloid: Petal-like; a sepal that resembles a petal.

petiole: A leaf stalk.

petiolules: The stalk of a leaflet of a compound leaf.

phyllary: An involucral bract in the Asteraceae.

pilose: With soft, straight hairs.

pinna(e): A leaflet or a primary division of a pinnately-compound leaf.

pinnate: Branching on opposite sides of an axis.

pinnatifid: Lobed half the distance to the midvein, but not reaching the midvein.

pinnatisect: Pinnately lobed to the midvein.

pistil: The female reproductive structure of a flower; stigma, style, and ovary.

pistillate: A flower with one or more pistils but no functional stamens; a female flower or plant.

polygamous: Bearing bisexual and unisexual flowers on the same plant.

poricidal: Opening by pores.

prickle: A spine-like extension from the epidermis of a stem or leaf; borne at irregular locations.

prostrate: Lying flat on the ground.

pubescence: With short, soft hairs.

pubescent: Covered with soft hairs.

pulvinus: A swelling at the base of a petiole.

punctate: Marked with dots, depressions, or glands.

pustulate: With a small swelling at the base of a hair.

raceme: A simple elongated inflorescence with stalked flowers; the ordering of flowering usually from the base to the apex.

rachis: An axis that bears flowers or leaflets.

radiate head: With both ray and disk florets in a head; e.g., some members of the Asteraceae.

ray floret: A strap-shaped or ligulate flower in the Asteraceae.

receptacle: Where the flowers are attached in the head of the Asteraceae.

recurved: Curved backward or downward.

reflexed: Abruptly bent downward or backward.

retrorse: Pointed downward or backward.

revolute: Rolled backward.

rhizome: A creeping, underground stem bearing scale-like leaves.

rugulose: Slightly wrinkled.

sagittate: Arrowhead-shaped.

scabrous: Rough to the touch like sandpaper; with short, bristly hairs.

scandent: Climbing in any manner.

scape: A flowering stem that is essentially leafless.

scarious: Thin, dry, membranous, but not green.

schizocarp: A dry fruit that separates into one-seeded segments at maturity.

scorpioid cyme: A determinate, unilateral, coiled inflorescence, resembling a scorpion's tail.

sepal: One of the outer leaf-like parts of a flower; usually green but sometimes colored.

septate: Divided into one or more partitions.

septum: A partition.

seral: A temporary stage in a plant community with its own characteristics that can remain for a very short time or for many years.

serrate: A saw-toothed leaf margin.

sessile: Not stalked.

sheath: A tubular structure surrounding a part.

shoot: The stem and its leaves.

simple (leaves): Of one piece, not compound; one blade per petiole.

spathe: A large bract subtending an inflorescence.

spatulate: Oblong with an attenuated base.

spicate: Resembling a spike.

sporocarp: A hard, nut-like structure containing sporangia.

stamen: Male reproductive structure of a flower; the filament and anther.

staminate: Bearing stamens; a male flower or a male plant.

stellate: Bearing branched, star-shaped hairs.

sterile: Infertile; a floret that does not produce a seed.

stigma: Portion of a pistil that receives pollen.

stipitate: Borne on a stalk.

stipules: A pair of appendages that are sometimes present at the point of attachment of the leaf to the stem.

strigose: With stiff, appressed hairs.

GLOSSARY

style: Portion of a pistil between the stigma and ovary.

sub-: Meaning slightly, almost, under, or almost.

subtend: Standing below or close to; a bract at the base of a flower.

subulate: Awl-shaped.

succulent: With thick, juicy parts.

superior ovary: With the sepals, petals, and stamens attached at or near the base of the ovary.

taproot: The main root axis from which smaller roots arise.

tepal: A segment of a perianth that is not differentiated into a calyx and corolla.

tendril: A segment of a stem or leaf modified into a slender, twining holdfast.

tomentose: Covered with short, matted, tangled or woolly hairs.

trifoliolate: A compound leaf bearing three leaflets.

trigonous: Three-sided.

tripinnately compound: Pinnately-compound three times.

truncate: Ending abruptly; cut almost squarely at the end.

tubercle: A small nodule.

tuberculate: Bearing tubercles.

turgid: Swollen.

umbel: An inflorescence with the pedicles arising at approximately the same point.

umbellate: Resembling an umbel; a flat-topped inflorescence whose pedicels arise from a common point.

unisexual: Of one sex; pistillate only or staminate only.

united: Joined together.

utricle: A small, one-seeded, inflated fruit.

vein: Conducting tissue; a vascular bundle.

vermiform: Worm-shaped.

verticel: A whorl.

vescicle: A small, bladder-like structure.

villous: With long, soft interlaced hairs.

viscid: Sticky.

whorl: Three or more leaves at a node.

wing: A thin, dry membranous expansion of an organ.

zygomorphic: With or approaching bilateral symmetry.

INDEX

INDEX

INDEX

INDEX

INDEX

N

O

P

INDEX

INDEX

LIST OF VERTEBRATES

Scientific and common names of birds are according to Rappole and Blacklock 1985; of mammals, Jones et al. 1992; of reptiles, Conant 1975.

BIRDS

Mourning dove *Zenaida macroura*
Northern bobwhite quail *Colinus virginianus*
Plain chachalaca *Ortalis vetula*
Rio Grande turkey *Melagris galopavo intermedia*
Sandhill crane *Grus canadensis*
Scaled quail *Callipepla squamata*
White-winged dove *Zenaida asiatica*

MAMMALS

Black-tailed jackrabbit *Lepus californica*
Collared peccary or javelina *Tayassu tajacu*
White-tailed deer *Odcoileus virginianus*

REPTILES

Texas tortoise *Gopherus berlandieri*